T0335573

VOLUME TWO HUNDRED AND EIGHT

ADVANCES IN
IMAGING AND
ELECTRON PHYSICS

EDITOR-IN-CHIEF

Peter W. Hawkes
CEMES-CNRS
Toulouse, France

VOLUME TWO HUNDRED AND EIGHT

ADVANCES IN
IMAGING AND
ELECTRON PHYSICS

Edited by

PETER W. HAWKES
CEMES-CNRS
Toulouse, France

Academic Press is an imprint of Elsevier
125 London Wall, London EC2Y 5AS, United Kingdom
525 B Street, Suite 1650, San Diego, CA 92101, United States
50 Hampshire Street, 5th Floor, Cambridge, MA 02139, United States
The Boulevard, Langford Lane, Kidlington, Oxford OX5 1GB, United Kingdom

ISBN: 978-0-12-815214-0
ISSN: 1076-5670

For information on all Academic Press publications
visit our website at https://www.elsevier.com/books-and-journals

Working together
to grow libraries in
developing countries

www.elsevier.com • www.bookaid.org

Publisher: Zoe Kruze
Acquisition Editor: Jason Mitchell
Editorial Project Manager: Peter Llewellyn
Production Project Manager: Divya KrishnaKumar
Designer: Alan Studholme

Typeset by VTeX

CONTENTS

CONTRIBUTORS

Bernhard J. Hoenders
Zernike Institute for Advanced Materials, University of Groningen, Groningen, The Netherlands

John A. Rouse
MEBS, 14 Cornwall Gardens, London, United Kingdom
Formerly Applied Optics Section, Blackett Laboratory, Imperial College of Science, Technology, and Medicine, London, United Kingdom

Dirk van Delft
Boerhaave Museum, Leiden, Netherlands

John van Gorkom (deceased)

Ton van Helvoort
Acta Biomedica, Elsloo, Netherlands

PREFACE

An account by Bernhard Hoenders of the fascinating phenomenon known as ghost imaging opens this volume. Here, an image seems to appear in mysterious circumstances and even turbulence can be tolerated. The experimental arrangements that can generate such images are described in detail together with the corresponding theory and the coexistence of classical and quantum theory to explain them is elucidated. Moreover, this is much more than a strange effect, it can benefit radiology and encryption, for example. A temporal form also exists in parallel with the spatial version. This lucid account of a new kind of image formation will, I am certain, be appreciated by readers to whom it is unknown or perplexing.

This is followed by a further (and last) chapter of the draft doctoral dissertation of the late John van Gorkom. The circumstances surrounding the writing and publication of this work are described at length by Ton van Helvoort and van Gorkom's supervisor Dirk van Delft in an appendix to the chapters published in volume 205 of these Advances. Here, van Gorkom describes the years that led to the decision by Siemens & Halske to embark on the commercial production of a transmission electron microscope. The frustrations and final success of the endeavors of Ernst Ruska and Bodo von Borries are recounted in detail and the role of Ernst Ruska's brother Helmut emerges clearly. (A full account of Helmut Ruska's participation is to be found in the article by Gelderblom & Krüger, 2014.) The activities of the scientists at AEG, primarily concerned with emission imaging, are also described but are considered less important during these years. An interesting foretaste of the scanning electron microscope is also given but the closing date is too early for van Gorkom to tell us about Manfred von Ardenne's instrument. Microscope developments in several European countries and in North America are charted, especially in Belgium where Ladislaus Marton built numerous microscopes. Readers will appreciate the thorough coverage of the literature of the 1930s, especially when the contents of recondite journals are recapitulated. One publication that van Gorkom would certainly have discussed slipped through the net: a paper by Martin (1934) in the *Journal of the Television Society* that preceded the article in *Science Progress* analyzed here. We also remind the reader of the first of the three reports on research at the AEG Research Institute (Ramsauer, 1941), which includes extensive information about work at

AEG from 1930 to 1940 – a total of 132 publications are listed by year; this was cited in the earlier article. The *Jahrbücher der AEG-Forschung* for the 1930s are also sources of complementary information and attention is drawn to the relevant volumes at the end of Ramsauer's compilation; only one article from these yearbooks was cited in the preceding article. A little more information has come to light concerning the Trillat–Fritz electron microscope, mentioned by van Gorkom in Section 5.3 and in his earlier publication (Hawkes, 2013). The place of Bodo von Borries in the story is the subject of a short (and little known) book by Götz von Borries (2001). A compilation published to mark the opening of the Siemens & Halske Laboratorium für Übermikroskopie is also of interest here (Siemens, 1941), though the contributors are primarily concerned with results obtained with the new Siemens instrument.

We conclude with a long article by John Rouse, reprinted from the *Advances in Optical and Electron Microscopy*. This is still the fullest account of the use of the finite-difference method of calculating potentials or field distributions in three dimensions. A complete account of the theory is given as well as several examples of the method. Rouse considers the spherical condenser, a bi-potential and a magnetic lens, lenses with elliptical bores, photomultiplier tubes, charging on insulating specimens in the scanning electron microscope (SEM), magnetic immersion lenses used in connection with lithography, Wien filters and focusing, deflection and collection systems in the SEM.

REFERENCES

Gelderblom, H. R., & Krüger, D. H. (2014). Helmut Ruska (1908–1973): His role in the evolution of electron microscopy in the life sciences and especially virology. *Advances in Imaging and Electron Physics, 182*, 1–94.

Hawkes, P. W. (2013). Trillat–Fritz: A very early French electron microscope. *In Focus, 31*, 24–27.

Martin, L. C. (1934). The paraxial equations of electron optics. *Journal of the Television Society, 1*, 377–387.

Ramsauer, C. (1941). *Zehn Jahre Elektronenmikroskopie, Ein Selbstbericht des AEG Forschungs-Instituts*. Berlin: Julius Springer.

Siemens (1941). *Das Übermikroskop als Forschungsmittel. Vorträge, gehalten anläßlich der Eröffnung des Laboratoriums für Übermikroskopie der Siemens & Halske A. G., Berlin*. Berlin: de Gruyter.

von Borries, G. (2001). *Bodo von Borries und das Elektronenmikroskop. Erfindung und Entwicklung*. Egelsbach: Fouqué.

Peter W. Hawkes

CHAPTER ONE

Review of a Bewildering Classical–Quantum Phenomenon: Ghost Imaging

B.J. Hoenders
Zernike Institute for Advanced Materials, University of Groningen, Groningen, The Netherlands
e-mail address: b.j.hoenders@rug.nl

Contents

1. INTRODUCTION

Ghost imaging is one of the recent fascinating and probably counter-intuitive topics of quantum optics.

Ghost–imaging experiments in general correlate the outputs from two photo detectors situated in two different paths: a high spatial resolution detector that measures a field which has not interacted with the object to be imaged, and a bucket (single pixel) detector that collects a field which has interacted with the object.

A variant of these classical experiments, known as "Computational Ghost Imaging", uses computed signals instead of using the signal generated by the field which has not interacted with the object.

Advances in Imaging and Electron Physics, Volume 208
ISSN 1076-5670
https://doi.org/10.1016/bs.aiep.2018.08.001

1

The name "ghost" comes from the fact that a field generated by an illuminated object in one path produces an image or interference fringes in another path, measuring coincident counting rates or second order correlations. The image or fringes depend on the intensities (fields) occurring in both paths.

Ghost imaging has opened up new perspectives to obtain highly resolved images, even if they are blurred by noise and turbulence.

For an early review see: Erkmen and Shapiro (2010). Recent reviews are: Moreau, Ermes, Thomas, and Padgett (2017), Padgett and Boyd (2017), and Shirai (2017). A very nice pedagogical introduction is given by Basano and Ottonello (2007).

Ghost Imaging was theoretically predicted by Belinskiĩ and Klyshko (1994) and Klyshko (1988a, 1988b).

The first ones to observe the "Ghost effect" are Belinskiĩ and Klyshko (1994). In their own words: "The following feature of two–photon devices is interesting to us at the applications level: if the illuminance in one beam (e.g., the upper one) is recorded without temporal selection, then it will be uniform and independent of the properties of the transparency in the other (lower) channel. Thus, the transformed information will be encoded, and the key to deciphering it is carried by the pulses from the lower photo detector, which can be transmitted by an independent communications channel, which enhances not only noise immunity, but also the degree of confidentiality of communication."

Ghost Imaging was demonstrated experimentally for the first time in 1995 by Pittman, Shih, Strekalov, and Sergienko (1995). Fig. 1 shows the experimental setup of the experiment performed by Pittman et al. (1995): A signal and idler photon pairs are produced in spontaneous parametric down-conversion. An aperture placed in front of a fixed detector is illuminated by the signal beam through a convex lens. A sharp magnified image of the aperture is found in the coincidence counting rate when a mobile detector is scanned in the transverse plane of the idler beam at a specific distance in relation to the lens.

Fig. 2 shows the result of the experiment performed by Pittman et al. (1995).

Ghost Interference and Diffraction is experimentally observed by Strekalov, Sergienko, Klyshko, and Shih (1995). They report diffraction and interference by two-photon correlation measurements. The signal and idler beams, produced by spontaneous parametric down-conversion, are sent in different directions, and detected by two distant pointlike photon

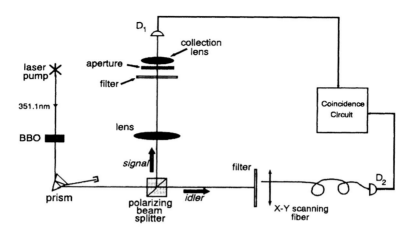

Figure 1 Cartoon schematic (not to scale) of the experimental setup.

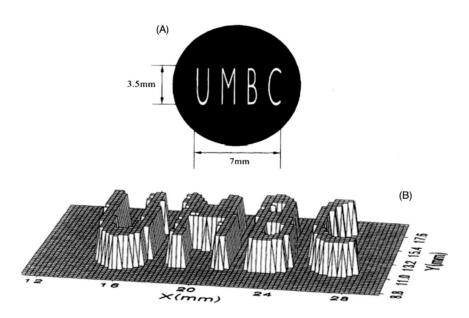

Figure 2 (A) Reproduction of the actual aperture placed in the signal beam. Note that the size of the letters is on the order of standard text. (B) Coincidence counts as a function of the fiber tips transverse plane coordinates. The scanning step size is 0.25 mm. The data shown is a "slice" at the half maximum value, with no image enhancement.

counting detectors. A double slit or a single slit is inserted into the signal beam. Then, interference–diffraction patterns are observed in coincidences by scanning the detector in the idler beam.

Fonseca, Souto Ribeiro, Pádua, and Monken (1999) discuss an interesting consequence of momentum entanglement: the ability of a two-photon field to mimic the scattering by a double slit, when this field is scattered by two spatially separated apertures, none of them being a double slit. Rather, the superposition of the two apertures do define a double slit, which determines the shape of the two-point fourth-order transverse spatial correlation function.

Hybrid schemes, using both diffraction and imaging, allow heralded phase-contrast imaging to be performed through introducing a phase filter in a Fourier plane of the crystal, see Aspden, Morris, He, Chen, and Padgett (2016). They utilize the position and orbital angular momentum correlations between signal and idler photons in order to obtain ghost images of a phase object.

This imaging technique enables imaging of phase objects using significantly fewer photons than standard phase-contrast imaging techniques.

The first experimental demonstrations of Ghost Imaging and Diffraction utilized entangled photon pairs generated by spontaneous parametric downconversion. This phenomenon was hence naturally ascribed to the quantum entanglement of photons. Furthermore, Abouraddy, Saleh, Sergienko, and Teich (2001) showed that the use of entangled photons in an imaging system can exhibit effects that cannot be mimicked by any other two-photon source, whatever the strength of the correlations between the two photons.

They consider a two-photon imaging system in which one photon is used to probe a remote (transmissive or scattering) object, while the other serves as a reference. Then the role of entanglement versus correlation is discussed in such a setting. Then, entanglement is shown to be a prerequisite for achieving distributed quantum imaging.

However, Bennink, Bentley, and Boyd (2002) show, both theoretically and experimentally, that ghost imaging can be realized with classical, mutually correlated beams. They also show that any kind of coincidence imaging technique which uses a bucket detector in the test arm is incapable of imaging phase-only objects, whether a classical or quantum source is employed.

Soon after this experimental work, the possibility of ghost imaging with classical thermal light was examined theoretically by Gatti, Brambilla, Bache, and Lugiato (2004a, 2004b). This theory was verified experimentally by use of pseudothermal light generated by a rotating ground glass plate and a laser beam (Ferri et al., 2005; Gatti et al., 2006).

Cai and Zhu (2004, 2005), and Cheng and Han (2004) show also that classical coherence theory suffices to describe ghost imaging and diffraction.

So it depends on the particular experiment whether or not a quantum description or classical description can- or has to be used.

Note that, although those experimental studies demonstrate that quantum entanglement is not necessary for ghost imaging, a similar experimental study leads to a different conclusion (Valencia, Scarcelli, D'Angelo, & Shih, 2005). Furthermore, Scarcelli, Berardi, and Shih (2006) claim in a subsequent study that ghost imaging is a two-photon correlation phenomenon that must be described quantum mechanically, irrespective whether or not the source is a classical or quantum source.

All these claims give rise to a lively dispute concerning the real physics of ghost imaging, see Gatti, Bondani, Lugiato, Paris, and Fabre (2007), Meyers, Deacon, and Shih (2008), and Scarcelli, Berardi, and Shih (2007). These discussions seem to be settled by the analysis based on the Gaussian states of the quantized electromagnetic field by Erkmen and Shapiro (2008). Furthermore, a unified theoretical treatment by Wang, Qamar, Zhu, and Zubairy (2009), describing both ghost imaging and the Hanbury Brown Twiss effect, shows that the application of classical physics suffices to describe both ghost imaging and the Hanbury Brown Twiss effect.

A surprising new form of ghost imaging is introduced by Shapiro (2008), who introduces the so-called computational ghost imaging. He demonstrates that ghost imaging with pseudothermal light can be fully described within the framework of classical optics. The experimental setup consists in particular of a computer-controlled spatial light modulator replacing a rotating ground glass plate for generating pseudothermal light and a computer calculated reference beam. The intensity correlations of these two beams are then measured.

This experiment is also performed by Bromberg, Katz, and Silberberg (2009). They demonstrate pseudothermal ghost imaging and ghost diffraction using only *a single detector*. This has been achieved replacing the high-resolution detector of the reference beam with a computation of the propagating field, following the proposal by Shapiro (2008). Since only a single detector has been used by Bromberg et al. (2009), this experiment provides experimental evidence that pseudothermal ghost imaging does ***not*** rely on nonlocal quantum correlations. Also the depth-resolving capability of this ghost imaging technique is shown.

The whole experiment is, as it turned out, equivalent with single pixel imaging, known in the field of signal processing (Duarte et al., 2008).

It has been shown by Aßmann and Bayer (2013) that the calculation times involved in the process of Ghost Imaging, considerably can be reduced using the technique of "Compressive adaptive computational ghost imaging". They propose a variant of previous techniques which are particularly well suited for computational ghost imaging.

A further interesting application of computational ghost imaging was analyzed by Wang, Zhao, Cheng, Gong, and Chen (2016). It is concerned with an optical image hiding scheme, having strong robustness and high security, which is based on computational ghost imaging. They propose a hiding scheme for a watermark encrypted with a computational ghost imaging system. Random speckle patterns compose a secret key.

It is now widely accepted that ghost imaging with pseudothermal light cannot be regarded as a quantum effect but has to be regarded as a fully classical effect. But, ghost imaging with entangled photons, generated by spontaneous parametric downconversion is essentially a manifestation of quantum effects.

A detailed analysis of the various concepts involved in ghost imaging theory like classical and quantum states of light; semiclassical versus quantum photodetection; and coherence propagation for phase-insensitive and phase-sensitive sources is contained in the review by Shapiro and Boyd (2012). They also clearly show (p. 975) the flaw in the early arguments explaining the necessity of the use of entangled states in order to make ghost imaging to work.

An "in depth" discussion is given by Ragy and Adesso (2011), who take an information theoretical approach for the analysis of the problem. They investigate the nature of correlations in Gaussian light sources used for ghost imaging, and adopt methods from quantum information theory to distinguish genuinely quantum- from classical correlations.

Combining a microscopic analysis of speckle–speckle correlations with an effective coarse-grained description of the beams, it is shown that quantum correlations exist even in classical like thermal light sources, a result which appears relevant for the implementation of ghost imaging in the regime of low illumination. The total correlations in the thermal source beams effectively determine the quality of the imaging, as quantified by the signal-to-noise ratio. See also Henderson and Vedral (2001), explaining in detail the relation between this problem and information theory.

A discussion of the so-called sparseness problem is given by Morris, Aspden, Bell, Boys, and Padgett (2015). They show that a ghost-imaging configuration is a useful approach for obtaining images with high signal-to-noise ratios by applying principles of image compression and associated image reconstruction to raw data formed from an average of fewer than one detected photon. The use of heralded single photons ensures that the background counts can be virtually eliminated from the recorded images.

Two color ghost imaging, i.e. ghost imaging with signal- and idler beams having different wavelengths, is described in an interesting experiment by Yong and Fu (2016). They showed that either ghost imaging or ghost diffraction can be achieved by suitable choices of two wavelengths of an illuminating source. A practical achievement made feasible by this technique is e.g. the possibility to have one beam with infrared wavelength to propagate through a medium and having a reference beam with a wavelength lying in the visible. The beam with infrared wavelength is not scattered as much if its wavelength were e.g. in the visible, and the signal beam can be detected with a better accuracy than if its wavelength had been in the infrared.

Chan, O'Sullivan, and Boyd (2009) study theoretical non-degenerate-wavelength (or two-color) ghost imaging using either thermal or quantum light sources. It is demonstrated that a high-quality ghost image can be obtained even when the wavelengths of light, used in the object and reference arms, are very different. In general the spatial resolution of the ghost image is found to depend on each of these wavelengths.

They also derive the counterintuitive result that resolution of non-degenerate-wavelength thermal ghost imaging can be higher than that of its quantum counterpart, although the photons have the same degree of spatial correlation in the two cases.

Duan, Du, and Xia (2013) analyze the multi-wavelength thermal ghost imaging using either a spatial light modulating system (SLMS) or a rotating ground glass plate (RGGP), respectively. It is found that the multi-wavelength ghost imaging dramatically enhances the signal-to-noise ratio (SNR) of ghost imaging for all types of objects.

The N-wavelength ghost imaging with SLMS produces N^2 ghost images incoherently added together to form one ghost image whose SNR is N^2 times the one of single-wavelength ghost imaging. The N-wavelength ghost imaging with RGGP produces N ghost images incoherently added together to form one ghost image whose SNR is N times the one of single-wavelength ghost imaging. Moreover, the color ghost image obtained by the multi-wavelength ghost imaging is better than the monochromatic ghost image.

Cheng and Lin (2013) develop a unified theory in order to investigate both thermal ghost imaging and ghost diffraction through a turbulent atmosphere. A very simple general valid analytical imaging formula is derived which contains several previous known results as particular cases.

Specifically, it is found that ghost imaging is more robust against turbulence than ghost diffraction. A comparison of ghost imaging with a conventional lens imaging system is given.

A totally different approach, based on similar ideas as employed in the understanding of ghost imaging, is the basis of an experiment where the photons which are used to illuminate the object do not have to be detected at all and no coincidence detection is necessary in order to achieve imaging (Lemos, Borish, Ramelow, Lapkiewicz, & Zeilinger, 2014).

The crux of this method is the use of two downconversion crystals which are so arranged that the are pumped by the same source. This then leads to two correlated beams. Following the comments of Moreau et al. (2017):

"Placing an object in the idler path between the two downconverted photon sources imposes a spatial variation on the distinguishability of the idler source and hence a spatial variation on the routing of the signal photons at the beam splitter. Hence the spatial variation of the object as illuminated by the idler is mapped to the signal even though the idler itself is never detected. In this system, the only detector that is required is an imaging array sensitive to the signal wavelength.

Despite this seemingly quantum process, as for quantum ghost imaging itself, it has been suggested that this scheme also has a classical counterpart with which it shares a number of features, see Shapiro and Venkatraman (2015)."

2. DESCRIPTION OF GHOST IMAGING WITH QUANTUM MECHANICAL AND CLASSICAL LIGHT

We will give two possible descriptions of ghost imaging, one generally applicable for both quantum- and classical light, the other for classical light only.

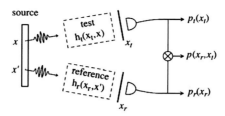

Figure 3 Two-photon coincidence imaging. The transfer function of the test system is to be obtained from the joint detection statistics using knowledge of the reference system.

2.1 Relations Between the Quantum and Classical Theory of Ghost Imaging

In this subsection we will show the parallels and differences between the quantum- and classical description of ghost imaging, as shown by Bennink et al. (2002), see also Abouraddy et al. (2001).

The theoretical scheme of ghost imaging is drawn in Fig. 3 (picture taken from Abouraddy et al., 2001 and Bennink et al., 2002).

The meaning of the symbols is: $p_r(x_r)$ denotes the single photon detection probability, $\bar{p}_r(x_r) = \int p(x_r, x_t) dx_t$ denotes the so-called marginal distribution, where $p(x_r, x_t)$ denotes joint probability detecting at x_t in the test arm, while detecting in coincidence a photon at x_r in the reference arm.

The two photon coincidence imaging relies completely on the fact that $p_r(x_r)$ and $\bar{p}_r(x_r)$ **can be different**. However classical probability theory leads to the result that they have the same value, provided that the integration over x_t (the bucket detector), covers all outcomes. Hence, a difference between these two outcomes is a violation of classical probability and therefore two photon coincidence imaging is an intrinsically quantum mechanical phenomenon. This conclusion was reached by Abouraddy et al. (2001).

The only way the total number of photons detected at x_r in conjunction with photons all over the test detector, is the loss of photons when they are traveling, i.e. if the test region is lossy, which means, in quantum mechanical terms, that $h_t(x_r)$ is non-unitary. Only then $p_r(x_r)$ and $\bar{p}_r(x_r)$ are different. A rigorous proof is given in Bennink et al. (2002) and runs as follows:

A two state quantum source is described by the state function $|\Psi\rangle = \int dx dx' \phi(x, x') |l_x, l_{x'}\rangle$, where $|l_x\rangle$ denotes a single quantum excitation at the point x on the source. The projections of these states onto the localized excitations at the test/reference detector are:

$$\langle l_{x_t} | \Psi \rangle = \int dx dx' h_t(x_t, x) \phi(x, x') | l_{x'} \rangle, \tag{1}$$

$$\langle l_{x_r} | \Psi \rangle = \int dx dx' h_r(x_r, x') \phi(x, x') | l_x \rangle, \tag{2}$$

$$\langle l_{x_i}, l_{x_r} | \Psi \rangle = \int dx dx' h_t(x_t, x) h_r(x_r, x') \phi(x, x'). \tag{3}$$

This leads to the single- and joint photon distributions:

$$p_t(x_t) = \int dx' \left| \int dx h_t(x_t, x) \varphi(x, x') \right|^2, \tag{4}$$

$$p_r(x_r) = \int dx' \left| \int dx h_r(x_r, x') \varphi(x, x') \right|^2, \tag{5}$$

$$p(x_t, x_r) = \left| \int \int dx' dx h_t(x_t, x) h_r(x_r, x') \phi(x, x') \right|^2, \tag{6}$$

where h_t and h_r are the impulse response functions of the reference and test systems, respectively. For a lossless system h_t is unitary, which means that $\int dx_t h_t(x_t, x) h_t^*(x_t, y) = \delta(x - y)$ holds.

Integrating Eq. (6) over x_t and substituting the unitarity property stated above leads to:

$$\bar{p}_r(x_r) = p_r(x_r). \tag{7}$$

Hence, two-photon coincidence imaging with a bucket detector cannot be used to image phase objects if the test system possesses loss or gain.

Next we will discuss the connection between the quantum mechanical and classical analysis of two photon coincidence imaging. A classical source will excite field distributions $E_n^t(x)$ and $E_n^r(x')$ in the test- and reference arms respectively, with a probability P_n. This leads to the following expected values for the various intensities:

$$p_t(x_t) = \sum_n P_n \left| \int dx h_t(x_t, x) E_n^t(x) \right|^2, \tag{8}$$

$$p_r(x_r) = \sum_n P_n \left| \int dx' h_r(x_r, x') E_n^r(x') \right|^2, \tag{9}$$

$$p_t(x_t, x_r) = \sum_n P_n \left| \int dx h_t(x_t, x) E_n^t(x) \right|^2 \left| \int dx' h_r(x_r, x') E_n^r(x') \right|^2. \tag{10}$$

If the classical probability distributions (8), (9), and (10) are required to be similar in form, then the following condition must be satisfied for the quantum probability distributions (4), (5), and (6):

$$\varphi(x, x')\varphi^*(y, y') = \sum_n P_n E_n^t(x) E_n^r(x') E_n^{*t}(y) E_n^{*r}(y'). \tag{11}$$

In case of a non-entangled source Eq. (11) is satisfied if $E_n^t(x)E_n^r(x') = \varphi(x, x')$, i.e. the classical source is a single state with unit probability. However, if the source is entangled, no such decomposition is possible.

Hence, a classical source cannot mimic a quantum source in a pure state for *all* test and reference systems, unless that state is not–entangled. However, this result does not imply that classical light sources cannot be used for ghost imaging. See Bennink et al. (2002). They show that entanglement is not required, and provide an experimental demonstration of coincidence imaging using a classical source.

The main condition for the applicability of classical sources for ghost imaging stems from the observation that only the condition (6) has to be satisfied. For a classical source this can be achieved as follows:

Suppose that the source field E_s^r generates a spot at the point s in the reference detector. Then $\left| \int dx' h_r(x_r, x') E_s^r(x') \right|^2 \approx \delta(x_r - s)$. Choosing $E_s^t = \alpha \int dx' h_t(s, x')\varphi(x, x')$, where $\alpha^{-2} \int ds = 1$, Eq. (10) becomes in the continuum limit ($P_s = \alpha^{-2}$ and $\sum_s P_s = \int ds P_s$):

$$p_t(x_t, x_r) = \sum_s P_s \alpha^2 \delta(x_r - s) \left| \int dx dx' h_t(x_t, x) h_r(s, x')\varphi(x, x') \right|^2$$

$$= \left| \int \int dx' dx h_t(x_t, x) h_r(x_r, x')\phi(x, x') \right|^2. \tag{12}$$

It is readily observed that (12) is the same as Eq. (6), hence the classical joint probability distribution is equal to the quantum joint probability distribution! Therefore, all phenomena depending only on this distribution, involving a test arm with known transfer function, can also be realized classically.

2.2 Analysis of Ghost Imaging with Classical Light

Basic setups for lensless ghost imaging and diffraction with classical thermal light is considered by Basano and Ottonello (2006), Chen, Liu, Luo, and Wu (2009), and Scarcelli et al. (2006).

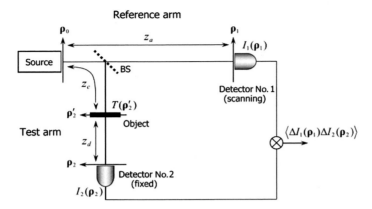

Figure 4 Conventional geometry for lensless Ghost Imaging with classical incoherent light. Spatially incoherent light from the source is split into two arms, called the test and the reference arms. An object (amplitude transmittance T) to be imaged is located in the test arm. The intensities are measured by two pointlike detectors located in each arm and the correlation between these two intensities is examined. Detector No. 1 in the reference arm is scanned to obtain information about the spatial profile of the object, while detector No. 2 is kept fixed in the test arm.

A scheme illustrating such an experimental setup is given in Fig. 4 (Shirai, 2017; Shirai, Setälä, & Friberg, 2011).

To simplify the analysis the vector nature of light is ignored, and the field is treated as a scalar quantity. It will be assumed that the scalar approximation holds if the object under consideration is insensitive to polarization.

We follow the theoretical analysis given by Shirai (2017). A light beam from a classical thermal source is split by a beam splitter (BS) into two beams. The field generated by the beam in the reference arm is detected, after free space propagation, by a scanning pointlike single-pixel detector 1 located at a point specified by a two-dimensional vector $\boldsymbol{\rho}_1$.

The beam in the test arm is incident on the object and the transmitted light is detected by a fixed single-pixel detector 2, located at a point specified by a two-dimensional vector $\boldsymbol{\rho}_2$. We assume that detector 2 is also pointlike for the moment. The effect of an extended active area of detector 2 on the ghost image and diffraction pattern will also be examined.

In the first experimental demonstration, the photon coincidence counting was performed to produce a ghost image and diffraction pattern (Pittman et al., 1995), whereas in the configuration under consideration the intensity correlations between the two detectors must be employed.

The intensity correlations $\langle I_1(\boldsymbol{\rho}_1;\omega)I_2(\boldsymbol{\rho}_2;\omega)\rangle$ are most conveniently expressed in terms of the intensity fluctuations $\Delta I_i(\boldsymbol{\rho}_i) = I_i(\boldsymbol{\rho}_i) - \langle I_i(\boldsymbol{\rho}_i)\rangle$:

$$\langle I_1(\boldsymbol{\rho}_1;\omega)I_2(\boldsymbol{\rho}_2;\omega)\rangle = \langle I_1(\boldsymbol{\rho}_1;\omega)\rangle\langle I_2(\boldsymbol{\rho}_2;\omega)\rangle + \langle \Delta I_1(\boldsymbol{\rho}_1;\omega)\Delta I_2(\boldsymbol{\rho}_2;\omega)\rangle,$$
(13)

where $I_1(\boldsymbol{\rho}_1;\omega)$ and $I_2(\boldsymbol{\rho}_1;\omega)$ are the instantaneous intensities at a point $\boldsymbol{\rho}_1$ in the reference arm and at a point $\boldsymbol{\rho}_2$ in the test arm, respectively, and the angular brackets denote the ensemble average.

Assuming Gaussian statistics, the moment theorem, also known as the Reed–Isserlis–Wick–Gauß moment theorem (Mandel & Wolf, 1995) leads to:

$$\langle \Delta I_1(\boldsymbol{\rho}_1;\omega)\Delta I_2(\boldsymbol{\rho}_2;\omega)\rangle = |W_{12}(\boldsymbol{\rho}_1,\boldsymbol{\rho}_2;\omega)|^2 = |\langle E_1(\boldsymbol{\rho}_1;\omega)E_1(\boldsymbol{\rho}_1;\omega)\rangle|^2.$$
(14)

Here $E_1(\boldsymbol{\rho}_1,\omega)$ and $E_1(\boldsymbol{\rho}_2,\omega)$ denote the fields at points at detector 1 in the reference arm, and detector 2 in the test arm respectively.

It is of paramount interest to notice that the analysis of the phenomenon "Ghost Imaging" rests completely on the validity of Eq. (14). This equation in its turn, rests completely on the fundamental assumption of Gaussian statistics. Only in this case the Isserlis–Reed–Wick–Gauß moment theorem is valid, enabling us to express the moment of any order of the stochastic process into second order moments.

The fields at the detectors can be expressed by their propagation equations from the source:

$$W_{12}(\boldsymbol{\rho}_1,\boldsymbol{\rho}_2;\omega) = \int\int W_s(\boldsymbol{\rho}_0,\boldsymbol{\rho}_0';\omega)G_1^*(\boldsymbol{\rho}_1,\boldsymbol{\rho}_0;\omega)G_2(\boldsymbol{\rho}_2,\boldsymbol{\rho}_0';\omega)d\boldsymbol{\rho}_0 d\boldsymbol{\rho}_0',$$
(15)

where

$$W_{12}(\boldsymbol{\rho}_1,\boldsymbol{\rho}_2;\omega) = \langle U_1^*(\boldsymbol{\rho}_1;\omega)U_1(\boldsymbol{\rho}_2;\omega)\rangle$$
(16)

is the cross-spectral density function between the fields $U_1(\boldsymbol{\rho}_1;\omega)$ at a point $\boldsymbol{\rho}_1$ (at detector 1) in the reference arm and $U_2(\boldsymbol{\rho}_2;\omega)$ at a point $\boldsymbol{\rho}_1$ (at detector 2) in the test arm.

The Green functions $G_1(\boldsymbol{\rho}_1,\boldsymbol{\rho}_0;\omega)$ and $G_2(\boldsymbol{\rho}_2,\boldsymbol{\rho}_0';\omega)$ denote the Green propagation functions for the reference- and the test arm respectively, and $W_s(\boldsymbol{\rho}_0,\boldsymbol{\rho}_0';\omega)$ denotes the cross-spectral density function at the source. The integrations have to be extended over the area of the source.

If the source, like the thermal source, is incoherent, i.e. $W_s(\boldsymbol{\rho}_0, \boldsymbol{\rho}_0'; \omega) = I_0\delta(\boldsymbol{\rho}_0' - \boldsymbol{\rho}_0')$, where I_0 denotes some constant, we obtain:

$$W_{12}(\boldsymbol{\rho}_1, \boldsymbol{\rho}_2; \omega) = I_0 \int G_1^*(\boldsymbol{\rho}_1, \boldsymbol{\rho}_0; \omega) G_2(\boldsymbol{\rho}_2, \boldsymbol{\rho}_0; \omega) d\boldsymbol{\rho}_0. \tag{17}$$

This result should be compared with Eq. (12) of the general theory developed above!

The Green propagation functions G_1 and G_2 for the reference- and the test arm respectively arise in connection with free space paraxial wave propagation and they read as:

$$G_1(\boldsymbol{\rho}_1, \boldsymbol{\rho}_0; \omega) = \frac{ik}{2\pi z_a} \exp\left[i\frac{k}{2z_a}(\boldsymbol{\rho}_1 - \boldsymbol{\rho}_2)^2\right], \tag{18}$$

where $k = \omega/c$ denotes the wave number, and:

$$G_2(\boldsymbol{\rho}_2, \boldsymbol{\rho}_0'; \omega) = -\frac{k^2}{(2\pi)^2 z_c z_d} \int T(\boldsymbol{\rho}_2'; \omega)$$
$$\times \exp\left\{i\frac{k}{2}\left[\frac{(\boldsymbol{\rho}_2' - \boldsymbol{\rho}_0)^2}{z_c} + \frac{(\boldsymbol{\rho}_2 - \boldsymbol{\rho}_2')^2}{z_d}\right]\right\} d\boldsymbol{\rho}_2'. \tag{19}$$

Eq. (19) is obtained by means of transport of the field from the source to the object with transmittance $T(\boldsymbol{\rho}_2'; \omega)$, and from the object to the detector 2. The propagation distances z_a, z_c, and z_d are specified in Fig. 4.

Next, let us assume that $z_a = z_c$, then inserting (18) and (19) into (17) yields:

$$\langle \Delta I_1(\boldsymbol{\rho}_1; \omega) \Delta I_2(\boldsymbol{\rho}_2; \omega)\rangle = |W_{12}(\boldsymbol{\rho}_1, \boldsymbol{\rho}_2; \omega)|^2 = \left(\frac{I_0 k}{2\pi z_d}\right)^2 |T(\boldsymbol{\rho}_1; \omega)|^2. \tag{20}$$

The result (20) shows that the knowledge of the intensity correlations leads to reconstruction of the object.

A similar calculation, using the condition $z_a = z_c + z_d$, leads to:

$$\langle \Delta I_1(\boldsymbol{\rho}_1; \omega) \Delta I_2(\boldsymbol{\rho}_2; \omega)\rangle$$
$$= \left(\frac{I_0 k}{2\pi z_d}\right)^2 \left|\int T(\boldsymbol{\rho}_2'; \omega) \exp\left(-i\frac{k}{z_d}(\boldsymbol{\rho}_2 - \boldsymbol{\rho}_1) \cdot \boldsymbol{\rho}_2'\right) d\boldsymbol{\rho}_2'\right|^2. \tag{21}$$

This equation shows that, provided the condition $z_a = z_c + z_d$ is satisfied, the ghost imaging process reveals the diffraction pattern of the object.

Many experiments are performed using a bucket detector, rather than a point-like detector in the test arm. One could think of the bucket detector as a point detector integrating the signal over an effective area, replacing the "point" by an area. We will analyze the outcome of the experiments performed above in case such a replacement of the detector in the reference arm has taken place.

The intensity fluctuations measured by the bucket detector are:

$$\Delta I_2'(\omega) = \int_D I_2(\boldsymbol{\rho}_2; \omega) d\boldsymbol{\rho}_2, \tag{22}$$

where D denotes the effective range of the detector. The intensity correlation fluctuations then become:

$$\langle \Delta I_1(\boldsymbol{\rho}_1; \omega) \Delta I_2(\boldsymbol{\rho}_2; \omega) \rangle = \int \langle \Delta I_1(\boldsymbol{\rho}_1; \omega) \Delta I_2(\boldsymbol{\rho}_2; \omega) \rangle d\boldsymbol{\rho}_2. \tag{23}$$

This equation, in combination with the Ghost Image result (20) yields:

$$\langle \Delta I_1(\boldsymbol{\rho}_1; \omega) \Delta I_2(\boldsymbol{\rho}_2; \omega) \rangle = D \left(\frac{I_0 k}{2\pi z_d} \right)^2 |T(\boldsymbol{\rho}_1; \omega)|^2. \tag{24}$$

Ghost Imaging is therefore not influenced by the change of a point detector by a bucket detector, due to the observation that the basic formula (20) is not dependent om $\boldsymbol{\rho}_2$.

However, considering Eq. (21), we readily observe that the integration over $\boldsymbol{\rho}_2$ blurs the diffraction pattern of the object. The insertion of a bucket detector will therefore not lead to a diffraction pattern.

An important quantity to be assessed is the visibility of the ghost image. Eq. (13) shows that the image always contains background terms $\langle I_1(\boldsymbol{\rho}_1; \omega) \rangle \langle I_2(\boldsymbol{\rho}_2; \omega) \rangle$. For a measure of the visibility V one usually takes:

$$V = \frac{[\langle \Delta I_1(\boldsymbol{\rho}_1; \omega) \Delta I_2(\boldsymbol{\rho}_2; \omega) \rangle]_{\text{max}}}{[\langle I_1(\boldsymbol{\rho}_1; \omega) I_2(\boldsymbol{\rho}_2; \omega) \rangle]_{\text{max}}}. \tag{25}$$

This definition is e.g. used by Gatti et al. (2006), Gatti et al. (2004a), and is quite useful as it appears to be closely related to the fundamental signal-to-noise ratio (Gatti et al., 2006; Erkmen & Shapiro, 2008). Eq. (25) can be rewritten as:

$$V = \frac{[|W_{12}(\boldsymbol{\rho}_1, \boldsymbol{\rho}_2)|^2]_{\text{max}}}{\langle I_1(\boldsymbol{\rho}_1; \omega) \rangle \langle I_2(\boldsymbol{\rho}_2; \omega) \rangle + [|W_{12}(\boldsymbol{\rho}_1, \boldsymbol{\rho}_2)|^2]_{\text{max}}}$$

$$= \frac{[|W_{12}(\boldsymbol{\rho}_1, \boldsymbol{\rho}_2)|^2]_{\max}}{W_{11}(\boldsymbol{\rho}_1, \boldsymbol{\rho}_1; \omega) W_{22}(\boldsymbol{\rho}_2, \boldsymbol{\rho}_2; \omega) + [|W_{12}(\boldsymbol{\rho}_1, \boldsymbol{\rho}_2)|^2]_{\max}} \tag{26}$$

where $W_{11}(\boldsymbol{\rho}_1, \boldsymbol{\rho}_1; \omega) = \langle I_1(\boldsymbol{\rho}_1; \omega) \rangle$ and $W_2(\boldsymbol{\rho}_2, \boldsymbol{\rho}_2; \omega) = \langle I_2(\boldsymbol{\rho}_1; \omega) \rangle$.

The Schwartz inequality leads to:

$$|W_{12}(\boldsymbol{\rho}_1, \boldsymbol{\rho}_2)|^2 \le W_{11}(\boldsymbol{\rho}_1, \boldsymbol{\rho}_1) W_{22}(\boldsymbol{\rho}_2, \boldsymbol{\rho}_2). \tag{27}$$

The maximum value of V is therefore obtained if

$$\left[|W_{12}(\boldsymbol{\rho}_1, \boldsymbol{\rho}_2)|^2\right]_{\max} = W_{11}(\boldsymbol{\rho}_1, \boldsymbol{\rho}_1) W_{22}(\boldsymbol{\rho}_2, \boldsymbol{\rho}_2). \tag{28}$$

So the maximal value for V is 1/2.

Zhao et al. (2012) extend pseudo-thermal light ghost imaging to the area of remote imaging and propose a ghost imaging lidar system. The experimental results demonstrate that the real-space image of a target at about 1.0 km range with 20 mm resolution is achieved by Ghost Imaging via sparsity constraints (GISC) technique.

Next we would like to analyze the requirements needed in order to make classical Ghost Imaging to work. To this end we quote from Hartmann, Molitor, and Elsäßer (2015):

"The requirements for a suitable light source in a classical Ghost Imaging experiment are:

(i)   nonconstant intensity autocorrelations $g^2(\tau)$, i.e., a super-Poissonian photon distribution,

(ii) spatial correlations that are able to resolve a target object, i.e., a transverse coherence length smaller than the object size, and

(iii) temporal correlations adapted to the resolution of the intensity correlation detection method with the coherence time τ_C directly linked to the spectral distribution of the source by $\Delta \nu = \frac{1}{\tau_C} = \frac{c}{\lambda_c^2}$ with the speed of light c, the central wavelength λ_c and the spectral width $\Delta\lambda$ in terms of wave-length. It is to be stressed that the latter requirement is the reason why Ghost Imaging experiments with thermal light have always been restricted to narrowband spectra.

For intensity–intensity correlation detection, coincidence counting techniques such as Hanbury Brown Twiss experiments are most commonly exploited, exhibiting a limited time-resolution given by the bandwidth of the detectors. This is why, experimentally, one must use either a pseudo-thermal light source which temporal coherence time is given by the interplay of the coherence time τ_C of the impinging laser source and the angular frequency of the rotating diffuser or a narrowband filter applied to light generated by lamps or other natural thermal light sources." (See Hartmann et al., 2015 for the pertinent references.)

Hartmann et al. (2015) show for the first time that a fully incoherent light source, in the sense that incoherence is given in both first- and second-order correlation, can be exploited efficiently in Ghost Imaging experiments.

An interesting variant of the classical Ghost Imaging scheme is proposed by Kuhn, Hartmann, and Elsäßer (2016): The key idea they use is that instead of measuring correlations between the object and reference beams such as in standard Ghost Imaging schemes, the light of the two beams is superimposed. The photon-statistics analysis of this mixed light allows then to determine the photon number distribution as well as to calculate the central second-order correlation coefficient.

The performance of this photon-statistics based Ghost Imaging system with one single detector is investigated in terms of visibility and resolution. Finally, the knowledge of the complete photon-statistics allows easy access to higher correlation coefficients enabling perform here third- and fourth-order Ghost Imaging.

The ideas developed in Kuhn et al. (2016) are further developed in Hartmann, Kuhn, and Elsässer (2016) by the derivation of analytical expressions for the spatial correlations. This model explains deviating characteristic features in comparison to state-of-the-art Ghost Imaging detection, such as maximum and minimum ghost image signals as well as the visibility performance, and explains the higher order correlations, which have been experimentally obtained by using a *single* detector.

2.3 Polarization and Polarimetry

Ghost imaging in connection with polarization and ellipsometry also has been investigated. One of the results, obtained by Shirai, Kellock, Setälä, and Friberg (2011), is the conclusion that the maximum attainable value of visibility defined by Eq. (25) (again) does not exceed 1/2. Differently stated: A background term is unavoidable in Ghost Imaging with classical light as far as the intensity correlation (i.e., the correlation between total intensities detected by each detector) is examined.

Shi, Hu, and Wang (2014) consider the case when the object and its background have the same reflectivity or transmittance, in which case conventional Ghost Imaging is helpless in detecting the object from the background. An improvement is obtained using the polarization components of the reflected or transmitted light, which allows the imaging of object buried in the same reflectivity or transmittance background. Using a combination of intensity and polarization information enables a better distinction between the background and material objects.

Surprisingly, a completely different result is obtained if Ghost Imaging with entangled light is considered: The visibility can approach unity even when the photon coincidence counting is performed without background subtraction (Gatti et al., 2004a).

Hannonen, Friberg, and Setälä (2016) introduce a novel ghost reflection ellipsometer for a spectral characterization of homogeneous thin films and interfaces. The device makes use of a uniform, spatially incoherent, unpolarized light source with Gaussian statistics and of the detection of intensity correlations. Unlike traditional ellipsometers, no source or detector calibration and reference sample are needed. The method is also insensitive to instrumentation errors.

The ellipsometer which is presented here is the classical analog of a quantum twin-photon arrangement discussed earlier in the literature. However, the classical configuration is easier to implement and to use, because entangled photon pairs are not needed and appropriate light sources and detectors are readily available.

Hannonen, Friberg, and Setälä (2017) use this technique introducing a novel approach for the spectral characterization of inhomogeneous thin films and interfaces by means of an imaging ghost ellipsometer operating with classical light obeying Gaussian statistics.

It is shown that the device output in general provides the ellipsometric information associated with the fractional Fourier transforms of the sample's reflection coefficients, which in special cases reduce to the Fourier transforms or images. The method is insensitive to instrumentation errors and, unlike in traditional ellipsometry, no source or detector calibration is needed.

3. GHOST IMAGING AND PHASE OBJECTS

(See Section 3.3, Shirai, 2017.) Optical imaging is an indispensable tool for modern biomedical research. Biological specimens examined with microscopes are mostly transparent and they must be regarded as phase objects. Optical imaging of phase objects has so been the subject of great importance over many years (for a recent review, see Mir, Bhaduri, Wang, Zhu, & Popescu, 2012).

It is known that, in general, pure phase objects cannot be seen with an ordinary imaging technique. Special techniques are required to visualize them, such as the phase contrast method developed by Zernike (Born & Wolf, 1975, Section 8.6.3). The same applies to ghost imaging. Pure phase objects have not been observed until recently with a framework of classical ghost imaging. Instead, either ghost diffraction (Bache et al., 2006; Borghi, Gori, & Santarsiero, 2006; Zhang et al., 2007), or phase retrieval (Gong & Han, 2010), has been employed to obtain phase information about the object rather indirectly. Ghost Imaging is basically coherent imaging even though the object is illuminated by incoherent thermal light (Gatti et al., 2006).

It is thus expected that pure phase objects can be visualized directly if Fourier-plane filtering could be performed somehow with a framework of classical ghost imaging. This is achievable in accordance with a recent study by Shirai et al. (2011).

Phase objects require special techniques making them visible. Bache et al. (2006) show that coherent imaging with incoherent classical thermal light is able to produce the interference pattern of a pure phase object, thus providing the ultimate demonstration that entanglement is not needed to do coherent imaging with incoherent light, not even in the case of a pure phase object. The only evident advantage of using entangled light might be that of obtaining a better visibility.

Bache et al. (2006) show also a remarkable complementary between Ghost Imaging and the Hanbury Brown Twiss (HBT) experiment, they state that: There exists a clear complementarity between the ghost diffraction scheme and the HBT scheme. In the HBT scheme the presence of a certain degree of spatial coherence is the essential ingredient that permits us to extract some phase information. The information becomes more correct as the coherence increases.

Conversely, the ghost diffraction scheme works as a coherent imaging scheme only thanks to the spatial incoherence of the light. The more the light is incoherent, the better the information is reconstructed.

These results contradict what was indicated in the introduction of Abouraddy, Stone, Sergienko, Saleh, and Teich (2004), where the possibility of doing coherent imaging in a ghost imaging scheme employing split thermal light was ascribed to the presence of spatial coherence.

Borghi et al. (2006) obtain the reconstruction of the phase by measuring the field (instead of the intensity) correlation function. In particular, the determination of the phase of the correlation function is made particularly easy and robust by the use of a suitably modified Young interferometer.

Zhang et al. (2007) observe experimentally Fourier transform Ghost Imaging of both amplitude-only and pure phase objects using classical incoherent light at Fresnel distance by a lensless scheme. Their interesting scheme provides a route toward aberration free diffraction limited three-dimensional imaging, using classical incoherent thermal light. This proposal supposedly does not have the resolution and depth-of-field limitations of lens-based tomographic systems.

Wang, Cai, and Hoenders (2014) show, using ghost-imaging techniques, that it is possible to reduce or even to annihilate the influence of aberrations connected with an arbitrary optical system. They consider a ghost-imaging setup, which consists of two arms, each containing an optical system. The reduction cancellation of the aberrations of the total imaging system is achieved by manipulating the values of the aberrations in one arm of the optical system. The technique is then applied for the optimal reconstruction of a weak phase object, manipulating the values of the defocusing such that the "Scherzer defocus condition" is obtained.

Gong and Han (2010) report an imaging approach, based on ghost imaging, which reconstructs a pure-phase object or a complex-valued object. Their analytical results, which are backed up by numerical simulations, demonstrate that both the complex-valued object and its amplitude-dependent part can be separately and non-locally reconstructed.

Gan et al. (2011) show cloaking in ghost imaging with thermal light. They show experimentally that when two identical phase-modulated objects are placed at equal distances from the thermal light source, they are offset completely in Ghost Imaging. It is shown that this effect, due to phase-conjugation, does not exist in two-photon entanglement source.

Zhang et al. (2014) perform an experimental realization of lensless Ghost Imaging for a phase-only object with pseudo-thermal light, originally proposed by Gong and Han (2010). In contrast with conventional Ghost Imaging, their scheme involves the interference of two correlated fields and completes the non-local lensless spatial reconstruction of both amplitude and phase distributions in Ghost Imaging with thermal light.

Liu, Cheng, and Han (2007) show that Ghost Imaging is feasible even when the coherent length on the object plane is larger than the character size of the object, a result which seems to be contradicting to the results obtained by Gatti et al. (2006). The phenomenon is explained in Fourier space. The analysis indicates that the resolution of the far-field diffraction image is determined by the coherent length on the back-focal plane, and that the distribution of the wave vector of the illumination on the object determines the range of information of a pixel detector collecting in the signal arm. The condition for Ghost Imaging is formulated in terms of the coherence length in the back focal plane.

4. RESOLUTION LIMIT AND CONTRAST TO NOISE RATIO

For systems utilizing the classical correlations from a pseudothermal light source, is the resolution limited by the speckle size in the object plane. The transverse coherence radius of the speckle pattern is described by the Van Cittert–Zernike theorem, thereby allowing the spatial resolution of the pseudothermal ghost imaging system to be determined (Gatti, Magatti, & Ferri, 2008).

This coherence radius is found to be of the order $dx = \lambda L/a_0$, in which the radius dx is determined by the wavelength λ of the pump used to illuminate the diffuser. The radius of the beam in the diffuser plane equals a_0. L is the distance of the object plane from that of the diffuser plane, or source of the speckle pattern (Gatti et al., 2008; Shapiro, 2008; Moreau et al., 2017).

Bai, Yang, and Yu (2010) analyze the noise properties in a two-arm microscope Ghost Imaging system with classical thermal light. They conclude that the noise enhances when the aperture of the reference lens increases, though the corresponding resolution increases. However, it is possible to obtain both a good resolution and a low noise level moving the object away from the original plane.

Cheng, Han, and Yan (2006) theoretically investigate the resolution and the noise in Ghost Imaging with classical thermal light, and they discuss the effects from the spatial coherence of the source and the aperture in the imaging system.

Zhang, Gong, Shen, Huang, and Han (2009) demonstrate that by making use of the second-order correlation of light fields, a two-arm imaging scheme, in which the reference arm contains a lens with a large numerical aperture, is feasible for improving the resolution of a conventional lens-limited imaging system.

Ferri, Magatti, Sala, and Gatti (2008) show that the longitudinal coherence of a speckle beam in lensless thermal Ghost Imaging is the key ingredient for determining the visibility and the spatial resolution of the retrieved image. Furthermore, both theoretically and experimentally, it was shown that lensless Ghost Imaging with thermal light is fully interpretable in terms of classical statistical optics. The visibility and the resolution of the ghost image are determined by the longitudinal coherence of the speckle beam, and no quantum explanation is necessary.

A general theory for the various resolution limits is derived by D'Angelo, Valencia, Rubin, and Shih (2005). They show that ghost images produced by separable sources are subject to the standard statistical limitations. However, entangled states offer the possibility of overcoming such limitations. Imaging can, therefore, obtain its fundamental limit through the high spatial resolution and nonlocal behavior of entangled systems. They obtain their results formulating the different physical pictures behind entangled and separable states in terms of a set of inequalities derived from the historical argument of Einstein, Podolsky, and Rosen.

Chan, O'Sullivan, and Boyd (2010) compare the performance of high-order thermal Ghost Imaging with that of conventional (that is, lowest-order) thermal ghost imaging for different data processing methods. Particular attention is given to high-order thermal Ghost Imaging with background normalization and to conventional ghost imaging with background subtraction.

It is found *analytically* that the contrast-to-noise ratio (CNR) of the normalized high-order ghost image is inversely proportional to the square root of the number of transmitting pixels of the object and that no data processing procedure performs better than lowest-order ghost imaging with background subtraction.

Chen et al. (2010) realize high-visibility Nth-order Ghost Imaging with thermal light by recording only the intensities in two optical paths in a lensless setup. It is shown that the visibility is dramatically enhanced as the order N increases, but longer integration times are required owing to the increased fluctuations of higher-order intensity correlation functions.

Liu, Chen, Luo, Wu, and Wu (2009) give an "in depth" theoretical analysis concerning the role of higher-order correlations of intensities with regard to ghost imaging using thermal light. These quantities give a measure of pure correlations among photons corresponding to multiphoton bunching. The synchronous detection of the same light field by all reference detectors is discussed.

It is found that the enhanced high visibility of Nth-order Ghost Imaging is a consequence of the contribution of N-photon bunching. This contribution is not small but is equal to the sum of all contributions from $(N-1)$ photon bunching.

5. GHOST IMAGING AND TURBULENCE

Both Ghost Imaging theory and experiments are applied to the phenomenon of turbulence, which usually affects the resolution and visibility of an image in long-distance imaging.

Meyers et al. (2008) demonstrate the possibility to obtain a ghost image from the scattered or reflected light from an object, explicitly stating that their result could only be interpreted non-classically. However, this statement was criticized by Shapiro and Boyd (2012). Then so-called "Standoff sensing" for surveillance was generally considered as a very promising possible application of Ghost Imaging. For such an application it is of paramount importance that a bright source illuminates the object in order to obtain sufficient statistics such that classical theory becomes applicable.

As Standoff sensing is used mostly in the open air, turbulence will influence the Ghost Imaging process. The earliest attempt to analyze the effect of atmospheric turbulence on the resultant ghost image has been made by Cheng (2009). On the basis of classical coherence theory and with the help of the extended Huygens–Fresnel principle the influence of turbulence is analyzed.

Cheng (2009) shows that increasing the turbulence strength and propagation distance, or decreasing the source size, will increase the size of the Point Spread Function (PSF), and leads to degradation of the imaging quality.

These results, using a classical theory, are confirmed by Li, Wang, Pu, Zhu, and Rao (2010). They find that the larger the source's transverse size, the better the quality of the ghost imaging. When the turbulence strength, propagation length, and the source's coherent width increase, the quality of the Ghost Imaging decreases and the image will finally become a Gaussian shape.

A more detailed analysis is conducted by Hardy and Shapiro (2011), who derive expressions for the spatial resolution, image contrast, and signal-to-noise ratio of a system imaging rough-surfaced targets in *reflection* through long, turbulent optical paths. They show, on the basis of the Gaussian-state framework, that the effects of turbulence on the quality of the resultant ghost image in the reflective case are similar to those derived by Cheng (2009) in the transmissive case.

Furthermore, Meyers, Deacon, and Shih (2011, 2012) report rather surprising experimental results that the quality of the ghost image is not degraded even though heating elements inducing atmospheric turbulence are inserted in both arms. The source is operated under low flux conditions, and the correlation measurement is performed by photon coincidence counting. Then, using the theory developed in Meyers et al. (2008), it is shown theoretically that the influence of the two–photon interference nature of the ghost imaging is the primary cause of this turbulence-free effect.

These results are in very stark contrast with those obtained previously, see the discussion above, and other investigations addressing the problem were published.

The effects of turbulence on ghost imaging are thoroughly theoretically investigated by Chan et al. (2011), and both theoretically and experimentally by Dixon et al. (2011). Using entangled photons, Dixon et al. (2011) demonstrate that for a specific experimental configuration the effect of turbulence can be greatly diminished.

Both Chan et al. (2011) and Dixon et al. (2011) obtain analytical expressions for the resolution and visibility, and demonstrate a method of ameliorating the effects of turbulence on Ghost Imaging systems.

The influence of turbulence on remote sensing is furthermore analyzed by Erkmen (2012). His analysis comprises signal to noise calculations taking into account the influence of photodetector noise, background radiation, and turbulence. Erkmen claims that: "The results reveal some key performance differences between computational ghost imaging and conventional active imaging, and identifies scenarios in which theory predicts that the former will perform better than the latter."

Following an entirely different approach it is shown theoretically that turbulence free imaging is possible using a modified configuration for conventional ghost diffraction with classical light (Shirai, Kellock, Setälä, & Friberg, 2012). The key principle of the method is ghost diffraction with classical incoherent light, and it employs an electromagnetic source and polarizing beam splitters.

They show theoretically that the effect of a phase aberration, which is insensitive to the direction of polarization of the light, but which may be deterministic or random, is successfully canceled out with this method. The squared modulus of the Fourier transform of the object is then obtained from the correlations of intensity fluctuations.

6. ENCRYPTION

Wang et al. (2016) propose a novel image hiding scheme based on computational ghost imaging to have strong robustness and high security. By them it is stated:

"The watermark is encrypted using a computational ghost imaging system, and random speckle patterns compose a secret key. The experimental and simulation results show that an authorized user can get the watermark with the secret key. The proposed scheme is robust against the "salt and pepper" noise and image cropping degradations."

Tanha, Kheradmand, and Ahmadi-Kandjani (2012) propose two approaches for optical encryption based on computational ghost imaging. These methods have the capability of encoding ghost images reconstructed from gray-scale images and colored objects. They experimentally demonstrate their approaches under eavesdropping in two different setups, thereby proving the robustness and simplicity of these setups for encryption as compared with previous algorithms.

A technique for simultaneous fusion, imaging, and encryption of multiple objects using a single-pixel detector has been proposed by Dongfeng et al. (2017). Encoded multiplexing patterns are employed to illuminate multiple objects simultaneously. The mixed light reflected from the objects is detected by a single pixel detector. An iterative reconstruction method is used to restore the fused image by summing the multiplexed patterns and detected intensities. Next, clear images of the objects are recovered by decoding the fused image. Experimentally fused and multiple clear images are obtained by utilizing a single-pixel detector to collect the direct and indirect reflected light.

Wang, Wang, Li, Yang, and Wu (2014) remark that:

"The recovered image in Ghost Imaging contains an error term when the number of measurements M is limited. By iteratively calculating the high-order error term, the iterative Ghost Imaging approach reconstructs a better image, compared to one recovered using a traditional Ghost Imaging approach, without adding complexity. We first propose an experimental scheme, for which iterative Ghost Imaging can be realized, namely the narrowed point spread function and exponentially increased signal-to-noise ratio (SNR) are realized. The exponentially increasing SNR when implementing iterative Ghost Imaging results from the replacement of M with M^k. Thus, a perfect recovery of the unknown object is demonstrated with M slightly bigger than the number of speckles in a typical light field. Based on our theoretical framework from the angle of high-order correlation R_k, the two critical behaviors of the iterative coefficients α and the measurements M are derived and well explained."

Chen and Chen (2014) report how an object and multiple hidden marks can be simultaneously recovered by using only one rebuilt reference intensity sequence in Ghost Imaging. The reconstructed object can be clearly observed during the decoding, and multiple marks can be effectively hidden. This unique characteristic is established using random selections of pixels from each reference intensity pattern, which also guarantees high security.

Clemente, Durán, Torres-Company, Tajahuerce, and Lancis (2010) show how computational ghost imaging can be used to encrypt and transmit object information to a remote party. Important features, such as key compressibility and vulnerability to eavesdropping, are experimentally analyzed.

Zhu et al. (2018) consider an optical image encryption scheme with multiple light paths based on compressive Ghost Imaging. M random phase-only masks (POMs) are generated by means of logistic map algorithm, and these masks are then uploaded to the spatial light modulator. The collimated laser light is divided into several beams by beam splitters as it passes through the spatial light modulator, and the light beams illuminate the secret images, which are converted into sparse images by discrete wavelet transform.

Wu et al. (2016) state that they:

"Propose an optical multiple image encryption scheme based on computational ghost imaging with position multiplexing. In the encryption process, each plain image is encrypted into an intensity vector by using the computational ghost imaging with a different diffraction distance. The final cipher text is generated by superposing all the intensity vectors together. Different from common multiple image crypto systems, the cipher text in the proposed scheme is simply an intensity vector instead of a complex amplitude."

Kong et al. (2013) introduce a new correlation operator and present a scheme to manipulate the spatial correlation of two-beam pairs. The manipulation of the spatial correlation serves as the encoding. Only when implementing the decryption process the perfect ghost image of the object can be extracted.

Li et al. (2016) (quote):

"Propose a multiple-image encryption method that is based on a modified logistic map algorithm, compressive ghost imaging, and coordinate sampling. In the encryption process, random phase-only masks are first generated with the modified logistic map algorithm; multiple secret images are transformed in order to be made sparse by the 2-D discrete cosine transformation (DCT) operation and scrambled by different random sequences; the scrambled images are then grouped to one combined image with the help of the coordinate sampling matrices; finally, putting the combined image in the object plane of the compressive ghost imaging system, the cipher text will be obtained from the bucket detector and transferred to the receivers."

7. GHOST IMAGING AND X-RAY

In medical applications of X-ray imaging the dose the patient is exposed to has to be as minimal as possible. Even when the radiation exposure is very small, ghost imaging with X-rays is expected to provide a high-resolution image of the object.

Very recently, ghost imaging with X-rays has been demonstrated experimentally by Pelliccia, Rack, Scheel, Cantelli, and Paganin (2016). Synchrotron emission from an ultrarelativistic electron bunch, provides a natural thermal source of hard X-rays for the experiment. In the experiment, a thin silicon crystal was employed as a beam splitter using Laue diffraction. An ultrafast imaging camera was employed to capture both the test and reference beams simultaneously. The bucket detection in the test arm was performed computationally by integrating the intensity of the test beam over a certain area. As a consequence, the ghost image of a copper wire was successfully obtained on the basis of the true intensity correlation extracted from noisy experimental data.

Since the phase and the amplitude of the object can be fully retrieved, in principle, from the modulus of its Fourier transform (Fienup, 1982, 1987; McBride, O'Leary, & Allen, 2004; Miao, Sayre, & Chapman, 1998), the principle of ghost diffraction is applicable to phase imaging. Accordingly, ghost diffraction with X-rays is expected to work as an alternative to coherent X-ray phase imaging. Furthermore, the whole information about the diffraction pattern can be acquired by this principle, whereas the low-frequency components of the diffraction pattern are always missing due to a beam stop in traditional X-ray diffraction experiments.

In the experimental demonstration by Yu et al. (2016), a pseudothermal X-ray source consisting of a synchrotron hard X-ray beam and a moving porous gold film are employed. The setup has only one arm, as opposed to the conventional ghost diffraction setup. Time-series measurements are performed to imitate the function of the conventional dual-arm setup in this single-arm one. Specifically, two intensity measurements are performed when the sample is inserted into the beam and when the sample is removed from the beam within a duration in which the X-ray speckle pattern from the pseudothermal source do not change, and then their product was calculated.

These processes are then repeated to take an average over these products. (This method is introduced as a perfect beam splitter for X-rays does not exist.) As a consequence, the intensity correlation obtained in this way clearly shows a diffraction pattern of a slit array object. The amplitude and the phase of this object are also successfully retrieved from this ghost diffraction pattern.

A highly coherent X-ray source is not required, the method can be implemented with laboratory X-ray sources and it also provides a potential solution for lensless diffraction imaging with fermions, such as neutrons and electrons where intensive coherent sources usually are not available.

In contrast to Ghost Imaging with X-rays, it may be difficult to mitigate the radiation damage in the particular geometry used by Yu et al. (2016), since the bucket detection is not available there. However, ghost diffraction with X-rays has some different advantages, as discussed in the early theoretical study by Cheng and Han (2004). They observe that X-ray imaging requires coherent sources which are available but are inherently weak compared to the sources which can be used for X-ray Ghost Imaging.

In recent years, a lot of attention has been given to phase imaging using a coherent X-ray beam. Liu, Nelson, Holzner, Andrews, and Pianetta (2013) review recent advances in several synchrotron-based hard X-ray phase contrast imaging XPCI methods. In particular when synchrotron hard X-rays are employed, the favorable brightness, energy tunability, monochromatic characteristics, and penetration depth have dramatically enhanced the quality and variety of XPCI methods. These features permit detection of the phase shift associated with 3D geometry of relatively large samples in a non-destructive manner.

Challenges and key factors in methodological development are discussed, and biological and medical applications are presented.

For an earlier review, describing the advantages of using X-ray phase information and a review of various techniques studied for X-ray phase imaging. See Momose (2005).

Another example of ghost imaging without the use of electromagnetic waves we mention the possibility to demonstrate experimentally ghost imaging with atoms (Khakimov et al., 2016).

In this case one utilizes quantum correlations between atoms. Two atom beams are formed by correlated pairs of ultracold, metastable helium atoms. Higher-order Kapitza–Dirac scattering is used in order to generate a large number of correlated atom pairs, enabling the creation of a clear ghost image with submillimeter resolution.

Ghost Imaging with atoms demonstrates the complementarity between light and matter. Moreover, this technique has great potential to test fundamental concepts in quantum mechanics with massive particles, such as ghost interference, Einstein–Podolsky–Rosen entanglement, and Bell's inequalities, see Jack et al. (2009). They use parametric down-conversion as the two-photon light source and show for phase objects, with differently oriented edges a violation of a Bell-type inequality for an orbital angular momentum subspace, thereby unambiguously revealing the quantum nature of their Ghost Imaging arrangement.

8. GHOST IMAGING AND DIFFRACTION IN TIME DOMAIN

Ghost imaging in the spatial domain has been considered hitherto, however, the temporal counterparts have also been examined both theoretically and experimentally.

The theoretical base for temporal ghost imaging is led by Kolner (1994). He observes a beautiful duality between the equations that describe the paraxial diffraction of beams confined in space and the dispersion of narrow-band pulses in dielectrics. This duality naturally leads to the conclusion that a quadratic phase modulation in time is the analog of a thin lens in space. Therefore, by a suitable combination of dispersion and quadratic phase modulation (now a "time lens"), one can synthesize the time-domain analog of an imaging system.

Such a temporal-imaging system can magnify time waveforms in the same manner as conventional spatial-imaging systems magnify scenes. The principles of temporal imaging are developed and time-domain analogs are derived for the imaging condition, magnification ratio, and impulse response of a temporal-imaging system.

We further mention Chen, Li, Li, Shi, and Zeng (2013) using a chaotic laser as the light source in temporal Ghost Imaging. The resulting temporal Ghost Imaging is the convolution between the transmission function of the object and the temporal correlation functions of the chaotic laser. The simulation experiment, which uses a controllable switch on time, as objects, shows the effectiveness of the scheme.

Devaux, Moreau, Denis, and Lantz (2016a) state in their own words, "We present a very simple device, inspired by computational ghost imaging, that allows the retrieval of a single nonreproducible, periodic, or nonperiodic, temporal signal. The reconstruction is performed by a single-shot spatially multiplexed measurement of the spatial intensity correlations between computer-generated random images and the images, modulated by a temporal signal, recorded, and summed on a chip CMOS camera used with no temporal resolution. Our device allows the reconstruction of either a single temporal signal with monochrome images or wavelength-multiplexed signals with color images." For supplementary material see Devaux, Moreau, Denis, and Lantz (2016b).

Setälä, Shirai, and Friberg (2010) observe that the two–photon coincidence detection amplitude obeys equations of motion similar to those obeyed by the correlations of classical, partially coherent plane-wave pulses in linear dispersive media. From this observation they conjectured that certain quantum-mechanical, time-domain entanglement phenomena like the shaping of nonlocal temporal and spectral pulses could be mimicked by classical partially coherent light.

They consider temporal ghost imaging with long, classical plane-wave pulses having a short coherence time. The correlation between the intensity fluctuations at the ends of the two arms of the ghost-imaging scheme is then given by a fractional Fourier transform of the temporal object in the test arm. As an imaging element a temporal lens is used in the reference arm. Choosing the dispersion properties of the arms and the "focal length" of the lens suitably, the general result can be reduced to an ordinary Fourier transform as well as to an image of the object, providing respectively spectral and temporal information on the object. Temporal ghost imaging with non-stationary pulses has also been analyzed by Shirai, Setälä, and Friberg (2010). In their own words:

"The optical setup for the analysis is constructed in such a way that a temporal object to be imaged is located in the test arm and that the reference arm consists simply of some temporal optical elements. The initial light to be incident on these two arms is assumed to be temporally incoherent non-stationary pulsed light. We have shown that, when a certain condition is satisfied, the correlation between intensity fluctuations in these two arms gives basically the squared modulus of the object, but it is generally distorted by the effect of the incident pulse. However, the effect of the distortion arising from the incident pulse is expected to be negligible when the effective temporal width of the incident pulse is much longer than that of the object; and, hence, almost complete information about the object is retrieved by means of the intensity correlation measurements in this case."

An important difference between temporal ghost imaging with stationary light and that with non-stationary light is the condition for the formation of the temporal image. The condition for temporal ghost imaging with stationary light is no longer applicable to the non-stationary case.

Ghost interference with classical partially coherent light pulses has been considered by Torres-Company, Lajunen, Lancis, and Friberg (2008). They observe that the two-photon temporal coincidence detection amplitude obeys a pair of equations identical to those of classical partially coherent plane-wave pulses propagating in linear dispersive media. These equations are also the same as the paraxial Wolf equations, for both the two-photon spatial probability amplitude and the cross-spectral density function, which implies that a fourfold analogy between space and time, as well as between quantum entanglement and partial coherence exists.

These analogies lead to the demonstration of fourth-order temporal nonlocal interferences with classical pulses assuming Gaussian statistics. The classical temporal counterpart of the ghost diffraction phenomenon is shown as well.

The analysis suggests that some time-domain entanglement phenomena, hitherto considered as uniquely quantum, can be mimicked by conventional partially coherent light pulses.

Another experiment (Devaux, Huy, Denis, Lantz, & Moreau, 2017) employs speckle ghost imaging of a single non-reproducible temporal signal, using pseudo-thermal speckle light patterns and a single detector array without any temporal resolution. A set of speckle patterns is generated deterministically, multiplied by the temporal signal and time integrated in a single shot by the camera. The temporal information is then retrieved by computing the spatial intensity correlations between this time integrated image and each speckle pattern of the set.

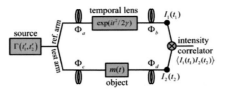

Figure 5 Illustration of the geometry for temporal ghost imaging. Light from a source is split into two mutually correlated beams which are directed to the reference and test arms. The reference arm contains two sections of single-mode fiber and a temporal lens characterized by the "focal length" γ. The test arm likewise contains two fiber sections and an object $m(t)$ between them. The fibers are characterized by the group-delay dispersion parameters Φ_i, with $i = (a, b, c, d)$. The output intensities of the arms, $I_i(t_i)$, with $i = (1, 2)$, are measured and correlated.

Denis, Moreau, Devaux, and Lantz (2017) use twin photons generated by spontaneous parametric down conversion to perform temporal ghost imaging of a single time signal. The experiment shows that temporal ghost imaging can be performed by using either twin photons or classical correlations, just as for spatial ghost imaging.

Ryczkowski, Barbier, Friberg, Dudley, and Genty (2016) performed Ghost Imaging experiments in time domain supporting the predictions made in the papers cited above.

In another paper, Ryczkowski, Barbier, Friberg, Dudley, and Genty (2017) propose and experimentally demonstrate a temporal Ghost Imaging scheme which generates a 5 times magnified ghost image of an ultrafast waveform. Inspired by shadow imaging in the spatial domain and building on the dispersive Fourier transform of an incoherent supercontinuum in an optical fiber, the approach overcomes the resolution limit of standard time–domain ghost imaging, generally imposed by the detectors speed.

9. THE THEORETICAL ANALYSIS OF TEMPORAL GHOST IMAGING

We follow the theoretical analysis given by Setälä et al. (2010). Fig. 5 depicts the experimental setup.

The slowly-varying envelopes of a realization of the random optical fields at the output of the arms, $E_i(t)$, can be written as:

$$E_i(t) = \int_{-\infty}^{+\infty} E_0(t')K_i(t, t')dt', \quad i = (1, 2), \tag{29}$$

where $E_0(t)$ is the input realization, and $i = (1, 2)$ refer to the reference arm and the test arm, respectively. Both arms consist of a cascade of linear elements whose effect on the field is contained in the kernels $K_{1,2}(t, t')$. For the reference arm we have:

$$K_1(t, t') = \frac{1}{2\pi} \sqrt{\frac{i}{\Phi_a}} \sqrt{\frac{i}{\Phi_b}} \int_{-\infty}^{+\infty} \exp\left(\frac{it''}{2\gamma}\right) \exp\left[-i\frac{(t''-t')^2}{2\Phi_a} - i\frac{(t''-t')^2}{2\Phi_b}\right] dt'', \tag{30}$$

whereas for the test arm we have:

$$K_2(t, t') = \frac{1}{2\pi} \sqrt{\frac{i}{\Phi_a}} \sqrt{\frac{i}{\Phi_b}} \int_{-\infty}^{+\infty} \exp\left(\frac{it''}{2\gamma}\right) \exp\left[-i\frac{(t''-t')^2}{2\Phi_c} - i\frac{(t''-t')^2}{2\Phi_d}\right] dt''. \tag{31}$$

The intensity correlation function of the output fields in the two arms then reads as:

$$G^{(2)}(t_1, t_2) = \langle I_1(t_1) I_2(t_2) \rangle$$
$$= \int_{-\infty}^{+\infty} \int_{-\infty}^{+\infty} \int_{-\infty}^{+\infty} \int_{-\infty}^{+\infty} \langle E_0^*(t_1') E_0^*(t_2') E_0^*(t_1'') E_0^*(t_2'')$$
$$\times K_1^*(t_1, t_1') K_2^*(t_2, t_2') K_1^*(t_1, t_1'') K_2(t_2, t_2'') \rangle dt_1' \, dt_2' \, dt_1'' \, dt_2''. \tag{32}$$

For fields obeying Gaussian statistics we can use the moment (Reed–Isserlis–Wick) theorem. Then, Eq. (32) becomes:

$$G^{(2)}(t_1, t_2) = \langle I_1(t_1) \rangle \langle I_2(t_2) \rangle + |\Gamma(t_1, t_2)|^2, \tag{33}$$

if

$$\Gamma(t_1, t_2) = \int_{-\infty}^{+\infty} \int_{-\infty}^{+\infty} \Gamma_0(t_1', t_2') K_1^*(t_1, t_1') K_2(t_2, t_2') dt_1'. \tag{34}$$

In this equation the function $\Gamma_0(t_1', t_2')$ denotes the correlation function of the incoming field, which in our case, assuming incoherent illumination, reads as:

$$\Gamma_0(t_1', t_2') = I_0 \delta(t_2' - t_1'). \tag{35}$$

Hence,

$$\Gamma(t_1, t_2) = I_0 \int_{-\infty}^{+\infty} \int_{-\infty}^{+\infty} K_1^*(t_1, t_1') K_2(t_2, t_2') dt_1', \tag{36}$$

which should be compared with (17) and (12), and contains, in a form of temporal correlation, the information of the object $m(t)$. Recalling (13) one readily obtains:

$$\langle \Delta I_1(t_1) \Delta I_2(t_2) \rangle = |\Gamma(t_1, t_2)|^2. \tag{37}$$

The function $\mathcal{M}_\alpha(\omega)$ denotes the Fourier fractional transform of order α, and reads as:

$$\mathcal{M}_\alpha(\omega) = \sqrt{\frac{1 - i \cot \alpha}{2\pi}} \exp(i\omega^2 \cot \alpha / 2)$$
$$\times \int_{-\infty}^{+\infty} m(t) \exp\left[-i\omega t \csc \alpha + it^2 \cot \alpha / 2\right] dt, \tag{38}$$

if

$$\cot \alpha = \frac{\Phi_b}{r} - \frac{1}{\Phi_d} - \frac{\gamma}{r} \quad \omega \csc \alpha = -\frac{t_2}{\Phi_d} - \frac{\gamma t_1}{r}, \tag{39}$$
$$r = (\Phi_a - \Phi_b)(\Phi_b - \gamma) - \Phi_b \gamma. \tag{40}$$

The final result then becomes:

$$|\Gamma(t_1, t_2)|^2 = \frac{I_0^2}{2\pi} \left| \frac{\gamma \sin \alpha}{r \Phi_d} \right| |\mathcal{M}_\alpha(\omega)|^2. \tag{41}$$

10. SPECTRAL CORRELATIONS AND GHOST IMAGING

Up to now we have considered ghost imaging resulting from spatial- and temporal correlated light. However, quoting Janassek, Blumenstein, and Elsäßer (2018), "another ghost modality (Scarcelli, Valencia, Gompers, & Shih, 2003), namely, ghost spectroscopy, has been developed that utilizes frequency quantum correlations of entangled photon pairs."

Paraphrasing their introduction: Specifically, twin photons from a spontaneous parametric down-conversion source are generated. Only the signal photons illuminate the sample and their totally transmitted light is registered by an integrating detector, whereas the idler photons are sent through a monochromator and subsequently recorded by a spectrally resolving detector. The frequency entanglement between the signal and idler photons leads to the reconstruction of the frequency-dependent absorption features of the sample when correlating the two detector signals as a function of the reference frequency.

Table 1 Introductory overview of representative experimental research on ghost-imaging modalities. The table summarizes representative achievements on GI, ghost spectroscopy, and temporal GI through quantum, thermal, and computational light-source approaches

Modality		Experimental demonstrations		
Terminology	Correlations	Twin photons	Thermal light	Computational
Ghost imaging	Spatial	Pitman et al. (1995)	Bennink et al. (2002), Ferri et al. (2005), Valencia et al. (2005)	Bromberg et al. (2009)
Ghost spectroscopy	Spectral	Scarcelli et al. (2003)	x	Multiplex Spectroscopy (Sloane, 1979)
Temporal GI	Temporal	Denis et al. (2017)	Devaux et al. (2017), Ryczkowski et al. (2016)	Devaux et al. (2016a)

Similar spooky quantum spectroscopy methods have already demonstrated real-world applications such as the determination of refractive indices and the absorption of CO_2 gas by Kalashnikov, Paterova, Kulik, and Krivitsky (2016).

Amiot, Ryczkowski, Friberg, Dudley, and Genty (2018) demonstrate a particular application of their ghost imaging technique (see Ryczkowski et al., 2017) to broadband spectroscopic measurements of methane absorption, and claim that their results offer novel perspectives for remote sensing in low light conditions, or in spectral regions where sensitive detectors are lacking.

For spectroscopy applications with low light illumination, an enhancement of the signal-to-noise performance has been achieved by multiplexing methods many years before the first Ghost Imaging experiment (Sloane, 1979).

Table 1, taken from Janassek et al. (2018), summarizes this brief overview of existing experimental realizations of Ghost imaging analogies by enumerating representative references. From there, it becomes clear that, even today, no ghost–spectroscopy experiment with classical light has been realized using thermal light sources. One of the reasons for this missing demonstration might be the technical demands of high time resolution τ_{meas} ($\tau_{meas} \ll \tau_c$) when intensity correlations of spectrally broadband light ($\tau_c = \frac{1}{\Delta \nu}$) are measured. A secondary reason could be the difficulty in finding a practical light source that provides a broadband spectrum, as well as the required wavelength correlations in order to enable the ghost-spectroscopy modality.

Janassek et al. (2018) then show experimentally that spectrally broadband light emitted by a quantum-dot superluminescent diode (SLD) indeed leads to high spectral resolution.

REFERENCES

Abouraddy, A. F., Saleh, B. E. A., Sergienko, A. V., & Teich, M. C. (2001). Role of entanglement in two-photon imaging. *Physical Review Letters, 87*, 123602.

Abouraddy, A. F., Stone, P. R., Sergienko, A. V., Saleh, B. E. A., & Teich, M. C. (2004). Entangled-photon imaging of a pure phase object. *Physical Review Letters, 93*, 213903.

Amiot, C., Ryczkowski, P., Friberg, A. T., Dudley, J. M., & Genty, G. (2018). Broadband continuous spectral ghost imaging for high resolution spectroscopy. arXiv:1805.12424.

Aspden, R. S., Morris, P. A., He, R., Chen, Q., & Padgett, M. J. (2016). Heralded phase-contrast imaging using an orbital angular momentum phase-filter. *Journal of Optics, 18*(5), 055204.

Aßmann, M., & Bayer, M. (2013). Compressive adaptive computational ghost imaging. *Scientific Reports, 3*, 1545.

Bache, M., Magatti, D., Ferri, F., Gatti, A., Brambilla, E., & Lugiato, L. A. (2006). Coherent imaging of a pure phase object with classical incoherent light. *Physical Review A, 73*, 053802.

Bai, Y., Yang, W., & Yu, X. (2010). Noise properties in a two-arm microscope imaging system with classical thermal light. *Applied Optics, 49*(24), 4554–4557.

Basano, L., & Ottonello, P. (2006). Experiment in lensless ghost imaging with thermal light. *Applied Physics Letters, 89*(9), 091109.

Basano, L., & Ottonello, P. (2007). Ghost imaging: Open secrets and puzzles for undergraduates. *American Journal of Physics, 75*(4), 343–351.

Belinskii, A. V., & Klyshko, D. N. (1994). Two-photon optics: Diffraction, holography, and transformation of two-dimensional signal. *Journal of Experimental and Theoretical Physics, 78*, 259–262.

Bennink, R. S., Bentley, S. J., & Boyd, R. W. (2002). "Two-photon" coincidence imaging with a classical source. *Physical Review Letters, 89*, 113601.

Borghi, R., Gori, F., & Santarsiero, M. (2006). Phase and amplitude retrieval in ghost diffraction from field-correlation measurements. *Physical Review Letters, 96*, 183901.

Born, M., & Wolf, E. (1975). *Principles of optics* (5th ed.). New York: Pergamon Press.

Bromberg, Y., Katz, O., & Silberberg, Y. (2009). Ghost imaging with a single detector. *Physical Review A, 79*, 053840.

Cai, Y., & Zhu, S.-Y. (2004). Ghost interference with partially coherent radiation. *Optics Letters, 29*(23), 2716–2718.

Cai, Y., & Zhu, S.-Y. (2005). Ghost imaging with incoherent and partially coherent light radiation. *Physical Review E, 71*, 056607.

Chan, K. W. C., O'Sullivan, M. N., & Boyd, R. W. (2009). Two-color ghost imaging. *Physical Review A, 79*, 033808.

Chan, K. W. C., O'Sullivan, M. N., & Boyd, R. W. (2010). Optimization of thermal ghost imaging: High-order correlations vs. background subtraction. *Optics Express, 18*(6), 5562–5573.

Chan, K. W. C., Simon, D. S., Sergienko, A. V., Hardy, N. D., Shapiro, J. H., Dixon, P. B., . . . Boyd, R. W. (2011). Theoretical analysis of quantum ghost imaging through turbulence. *Physical Review A*, *84*, 043807.

Chen, W., & Chen, X. (2014). Marked ghost imaging. *Applied Physics Letters*, *104*(25), 251109.

Chen, X.-H., Agafonov, I. N., Luo, K.-H., Liu, Q., Xian, R., Chekhova, M. V., & Wu, L.-A. (2010). High-visibility, high-order lensless ghost imaging with thermal light. *Optics Letters*, *35*(8), 1166–1168.

Chen, X.-H., Liu, Q., Luo, K.-H., & Wu, L.-A. (2009). Lensless ghost imaging with true thermal light. *Optics Letters*, *34*(5), 695–697.

Chen, Z., Li, H., Li, Y., Shi, J., & Zeng, G. (2013). Temporal ghost imaging with a chaotic laser. *Optical Engineering*, *52*, 076103.

Cheng, J. (2009). Ghost imaging through turbulent atmosphere. *Optics Express*, *17*(10), 7916–7921.

Cheng, J., & Han, S. (2004). Incoherent coincidence imaging and its applicability in X-ray diffraction. *Physical Review Letters*, *92*, 093903.

Cheng, J., Han, S.-S., & Yan, Y.-J. (2006). Resolution and noise in ghost imaging with classical thermal light. *Chinese Physics B*, *15*(9), 2002.

Cheng, J., & Lin, J. (2013). Unified theory of thermal ghost imaging and ghost diffraction through turbulent atmosphere. *Physical Review A*, *87*, 043810.

Clemente, P., Durán, V., Torres-Company, V., Tajahuerce, E., & Lancis, J. (2010). Optical encryption based on computational ghost imaging. *Optics Letters*, *35*(14), 2391–2393.

D'Angelo, M., Valencia, A., Rubin, M. H., & Shih, Y. (2005). Resolution of quantum and classical ghost imaging. *Physical Review A*, *72*, 013810.

Denis, S., Moreau, P.-A., Devaux, F., & Lantz, E. (2017). Temporal ghost imaging with twin photons. *Journal of Optics*, *19*(3), 034002.

Devaux, F., Huy, K. P., Denis, S., Lantz, E., & Moreau, P.-A. (2017). Temporal ghost imaging with pseudo-thermal speckle light. *Journal of Optics*, *19*(2), 024001.

Devaux, F., Moreau, P.-A., Denis, S., & Lantz, E. (2016a). Computational temporal ghost imaging. *Optica*, *3*(7), 698–701.

Devaux, F., Moreau, P.-A., Denis, S., & Lantz, E. (2016b). Computational temporal ghost imaging, supplementary material. *Optica*. https://doi.org/10.1364/optica.3.000698. s001.

Dixon, P. B., Howland, G. A., Chan, K. W. C., O'Sullivan-Hale, C., Rodenburg, B., Hardy, N. D., . . . Howell, J. C. (2011). Quantum ghost imaging through turbulence. *Physical Review A*, *83*, 051803.

Dongfeng, S., Jian, H., Yingjian, W., Kee, Y., Chenbo, X., Dong, L., & Wenyue, Z. (2017). Simultaneous fusion, imaging and encryption of multiple objects using a single-pixel detector. *Scientific Reports*, *7*, 13172.

Duan, D., Du, S., & Xia, Y. (2013). Multiwavelength ghost imaging. *Physical Review A*, *88*, 053842.

Duarte, M. F., Davenport, M. A., Takhar, D., Laska, J. N., Sun, T., Kelly, K. F., & Baraniuk, R. G. (2008). Single-pixel imaging via compressive sampling. *IEEE Signal Processing Magazine*, *25*(2), 83–91.

Erkmen, B. I. (2012). Computational ghost imaging for remote sensing. *Journal of the Optical Society of America A*, *29*(5), 782–789.

Erkmen, B. I., & Shapiro, J. H. (2008). Unified theory of ghost imaging with Gaußian-state light. *Physical Review A*, *77*, 043809.

Erkmen, B. I., & Shapiro, J. H. (2010). Ghost imaging: From quantum to classical to computational. *Advances in Optics and Photonics*, *2*(4), 405–450.

Ferri, F., Magatti, D., Gatti, A., Bache, M., Brambilla, E., & Lugiato, L. A. (2005). High-resolution ghost image and ghost diffraction experiments with thermal light. *Physical Review Letters*, *94*, 183602.

Ferri, F., Magatti, D., Sala, V. G., & Gatti, A. (2008). Longitudinal coherence in thermal ghost imaging. *Applied Physics Letters*, *92*(26), 261109.

Fienup, J. R. (1982). Phase retrieval algorithms: A comparison. *Applied Optics*, *21*(15), 2758–2769.

Fienup, J. R. (1987). Reconstruction of a complex-valued object from the modulus of its Fourier transform using a support constraint. *Journal of the Optical Society of America A*, *4*(1), 118–123.

Fonseca, E. J. S., Souto Ribeiro, P. H., Pádua, S., & Monken, C. H. (1999). Quantum interference by a nonlocal double slit. *Physical Review A*, *60*, 1530–1533.

Gan, S., Zhang, S.-H., Zhao, T., Xiong, J., Zhang, X., & Wang, K. (2011). Cloaking of a phase object in ghost imaging. *Applied Physics Letters*, *98*(11), 111102.

Gatti, A., Bache, M., Magatti, D., Brambilla, E., Ferri, F., & Lugiato, L. A. (2006). Coherent imaging with pseudo-thermal incoherent light. *Journal of Modern Optics*, *53*(5–6), 739–760.

Gatti, A., Bondani, M., Lugiato, L. A., Paris, M. G. A., & Fabre, C. (2007). Comment on "Can two-photon correlation of chaotic light be considered as correlation of intensity fluctuations?". *Physical Review Letters*, *98*, 039301.

Gatti, A., Brambilla, E., Bache, M., & Lugiato, L. A. (2004a). Correlated imaging, quantum and classical. *Physical Review A*, *70*, 013802.

Gatti, A., Brambilla, E., Bache, M., & Lugiato, L. A. (2004b). Ghost imaging with thermal light: Comparing entanglement and classical correlation. *Physical Review Letters*, *93*, 093602.

Gatti, A., Magatti, D., & Ferri, F. (2008). Three-dimensional coherence of light speckles: Theory. *Physical Review A*, *78*, 063806.

Gong, W., & Han, S. (2010). Phase-retrieval ghost imaging of complex-valued objects. *Physical Review A*, *82*, 023828.

Hannonen, A., Friberg, A. T., & Setälä, T. (2016). Classical spectral ghost ellipsometry. *Optics Letters*, *41*(21), 4943–4946.

Hannonen, A., Friberg, A. T., & Setälä, T. (2017). Classical ghost-imaging spectral ellipsometer. *Journal of the Optical Society of America A*, *34*(8), 1360–1368.

Hardy, N. D., & Shapiro, J. H. (2011). Reflective ghost imaging through turbulence. *Physical Review A*, *84*, 063824.

Hartmann, S., Kuhn, S., & Elsässer, W. (2016). Characteristic properties of the spatial correlations and visibility in mixed light ghost imaging. *Applied Optics*, *55*(28), 7972–7979.

Hartmann, S., Molitor, A., & Elsäßer, W. (2015). Ultrabroadband ghost imaging exploiting optoelectronic amplified spontaneous emission and two-photon detection. *Optics Letters*, *40*(24), 5770–5773.

Henderson, L., & Vedral, V. (2001). Classical, quantum and total correlations. *Journal of Physics. A. Mathematical and General*, *34*(35), 6899.

Jack, B., Leach, J., Romero, J., Franke-Arnold, S., Ritsch-Marte, M., Barnett, S. M., & Padgett, M. J. (2009). Holographic ghost imaging and the violation of a Bell inequality. *Physical Review Letters*, *103*, 083602.

Janassek, P., Blumenstein, S., & Elsäßer, W. (2018). Ghost spectroscopy with classical thermal light emitted by a superluminescent diode. *Physical Review Applied*, *9*, 021001.

Kalashnikov, D. A., Paterova, A. V., Kulik, S. P., & Krivitsky, L. A. (2016). Infrared spectroscopy with visible light. *Nature Photonics*, *10*(98), 98–101.

Khakimov, R. I., Henson, B. M., Shin, D. K., Hodgman, S. S., Dall, R. G., Baldwin, K. G. H., & Truscott, A. G. (2016). Ghost imaging with atoms. *Nature*, *540*, 100–103.

Klyshko, D. N. (1988a). A simple method of preparing pure states of an optical field, of implementing the Einstein–Podolsky–Rosen experiment, and of demonstrating the complementarity principle. *Soviet Physics. Uspekhi*, *31*(1), 74.

Klyshko, D. N. (1988b). Combine EPR and two-slit experiments: Interference of advanced waves. *Physics Letters A*, *132*(6), 299–304.

Kolner, B. H. (1994). Space–time duality and the theory of temporal imaging. *IEEE Journal of Quantum Electronics*, *30*(8), 1951–1963.

Kong, L.-J., Li, Y., Qian, S.-X., Li, S.-M., Tu, C., & Wang, H.-T. (2013). Encryption of ghost imaging. *Physical Review A*, *88*, 013852.

Kuhn, S., Hartmann, S., & Elsäßer, W. (2016). Photon-statistics-based classical ghost imaging with one single detector. *Optics Letters*, *41*(12), 2863–2866.

Lemos, G. B., Borish, V., Ramelow, S., Lapkiewicz, R., & Zeilinger, A. (2014). Quantum imaging with undetected photons. *Nature*, *512*, 409–416.

Li, C., Wang, T., Pu, J., Zhu, W., & Rao, R. (2010). Ghost imaging with partially coherent light radiation through turbulent atmosphere. *Applied Physics B*, *99*(3), 599–604.

Li, X., Meng, X., Yang, X., Yin, Y., Wang, Y., Peng, X., . . . Chen, H. (2016). Multiple-image encryption based on compressive ghost imaging and coordinate sampling. *IEEE Photonics Journal*, *8*(4), 1–11.

Liu, H., Cheng, J., & Han, S. (2007). Ghost imaging in Fourier space. *Journal of Applied Physics*, *102*(10), 103102.

Liu, Q., Chen, X.-H., Luo, K.-H., Wu, W., & Wu, L.-A. (2009). Role of multiphoton bunching in high-order ghost imaging with thermal light sources. *Physical Review A*, *79*, 053844.

Liu, Y., Nelson, J., Holzner, C., Andrews, J. C., & Pianetta, P. (2013). Recent advances in synchrotron-based hard x-ray phase contrast imaging. *Journal of Physics. D, Applied Physics*, *46*(49), 494001.

Mandel, L., & Wolf, E. (1995). *Optical coherence and quantum optics* (Chapter 1.6.1). New York: Cambridge University Press.

McBride, W., O'Leary, N. L., & Allen, L. J. (2004). Retrieval of a complex-valued object from its diffraction pattern. *Physical Review Letters*, *93*, 233902.

Meyers, R., Deacon, K. S., & Shih, Y. (2008). Ghost-imaging experiment by measuring reflected photons. *Physical Review A*, 77, 041801.

Meyers, R. E., Deacon, K. S., & Shih, Y. (2011). Turbulence-free ghost imaging. *Applied Physics Letters*, *98*(11), 111115.

Meyers, R. E., Deacon, K. S., & Shih, Y. (2012). Positive–negative turbulence-free ghost imaging. *Applied Physics Letters*, *100*(13), 131114.

Miao, J., Sayre, D., & Chapman, H. N. (1998). Phase retrieval from the magnitude of the Fourier transforms of nonperiodic objects. *Journal of the Optical Society of America A*, *15*(6), 1662–1669.

Mir, M., Bhaduri, B., Wang, R., Zhu, R., & Popescu, G. (2012). Quantitative phase imaging. In E. Wolf (Series Ed.), *Progress in optics: Vol. 57* (pp. 133–217). Elsevier.

Momose, A. (2005). Recent advances in x-ray phase imaging. *Japanese Journal of Applied Physics*, *44*(9R), 6355–6367.

Moreau, P.-A., Ermes, T., Thomas, G., & Padgett, M. J. (2017). Ghost imaging using optical correlations. *Laser & Photonics Reviews*, *12*(1), 1700143.

Morris, P. A., Aspden, R. S., Bell, J. E. C., Boys, R. W., & Padgett, M. J. (2015). Imaging with a small number of photons. *Nature Communications*, *6*(6), 5913.

Padgett, M. J., & Boyd, R. W. (2017). An introduction to ghost imaging: Quantum and classical. *Philosophical Transactions of the Royal Society of London A: Mathematical, Physical and Engineering Sciences*, *375*(2099).

Pelliccia, D., Rack, A., Scheel, M., Cantelli, V., & Paganin, D. M. (2016). Experimental X-ray ghost imaging. *Physical Review Letters*, *117*, 113902.

Pittman, T. B., Shih, Y. H., Strekalov, D. V., & Sergienko, A. V. (1995). Optical imaging by means of two-photon quantum entanglement. *Physical Review A*, *52*, R3429–R3432.

Ragy, S., & Adesso, G. (2011). Nature of light correlations in ghost imaging. *Scientific Reports*, *2*, 651.

Ryczkowski, P., Barbier, M., Friberg, A. T., Dudley, J. M., & Genty, G. (2016). Ghost imaging in the time domain. *Nature Photonics*, *10*, 167–170.

Ryczkowski, P., Barbier, M., Friberg, A. T., Dudley, J. M., & Genty, G. (2017). Magnified time-domain ghost imaging. *APL Photonics*, *2*(4), 046102.

Scarcelli, G., Berardi, V., & Shih, Y. (2006). Can two-photon correlation of chaotic light be considered as correlation of intensity fluctuations? *Physical Review Letters*, *96*, 063602.

Scarcelli, G., Berardi, V., & Shih, Y. H. (2007). Scarcelli, Berardi, and Shih reply. *Physical Review Letters*, *98*, 039302.

Scarcelli, G., Valencia, A., Gompers, S., & Shih, Y. (2003). Remote spectral measurement using entangled photons. *Applied Physics Letters*, *83*(26), 5560–5562.

Setälä, T., Shirai, T., & Friberg, A. T. (2010). Fractional Fourier transform in temporal ghost imaging with classical light. *Physical Review A*, *82*, 043813.

Shapiro, J. H. (2008). Computational ghost imaging. *Physical Review A*, *78*, 061802.

Shapiro, J. H., & Boyd, R. W. (2012). The physics of ghost imaging. *Quantum Information Processing*, *11*, 949–993.

Shapiro, J. H., & Venkatraman, D. (2015). Classical imaging with undetected photons. *Science Reports*, *5*, 10329.

Shi, D., Hu, S., & Wang, Y. (2014). Polarimetric ghost imaging. *Optics Letters*, *39*(5), 1231–1234.

Shirai, T. (2017). Modern aspects of intensity interferometry with classical light. In *Progress in optics: Vol. 62* (pp. 1–72). Elsevier.

Shirai, T., Kellock, H., Setälä, T., & Friberg, A. T. (2011). Visibility in ghost imaging with classical partially polarized electromagnetic beams. *Optics Letters*, *36*(15), 2880–2882.

Shirai, T., Kellock, H., Setälä, T., & Friberg, A. T. (2012). Imaging through an aberrating medium with classical ghost diffraction. *Journal of the Optical Society of America A*, *29*(7), 1288–1292.

Shirai, T., Setälä, T., & Friberg, A. T. (2010). Temporal ghost imaging with classical non-stationary pulsed light. *Journal of the Optical Society of America B*, *27*(12), 2549–2555.

Shirai, T., Setälä, T., & Friberg, A. T. (2011). Ghost imaging of phase objects with classical incoherent light. *Physical Review A*, *84*, 041801.

Sloane, N. J. (1979). Multiplexing methods in spectroscopy. *Mathematics Magazine*, *52*(2), 71–80.

Strekalov, D. V., Sergienko, A. V., Klyshko, D. N., & Shih, Y. H. (1995). Observation of two-photon "ghost" interference and diffraction. *Physical Review Letters*, *74*, 3600–3603.

Tanha, M., Kheradmand, R., & Ahmadi-Kandjani, S. (2012). Gray-scale and color optical encryption based on computational ghost imaging. *Applied Physics Letters*, *101*(10), 101108.

Torres-Company, V., Lajunen, H., Lancis, J., & Friberg, A. T. (2008). Ghost interference with classical partially coherent light pulses. *Physical Review A*, 77, 043811.

Valencia, A., Scarcelli, G., D'Angelo, M., & Shih, Y. (2005). Two-photon imaging with thermal light. *Physical Review Letters*, *94*, 063601.

Wang, F., Cai, Y., & Hoenders, B. J. (2014). Reduction or annihilation of aberrations of an optical system by balancing ghost-imaging technique and optimal imaging of a pure weak phase object. *Journal of the Optical Society of America A*, *31*(1), 48–57.

Wang, L., Zhao, S., Cheng, W., Gong, L., & Chen, H. (2016). Optical image hiding based on computational ghost imaging. *Optics Communications*, *366*, 314–320.

Wang, L.-G., Qamar, S., Zhu, S.-Y., & Zubairy, M. S. (2009). Hanbury Brown Twiss effect and thermal light ghost imaging: A unified approach. *Physical Review A*, *79*, 033835.

Wang, W., Wang, Y. P., Li, J., Yang, X., & Wu, Y. (2014). Iterative ghost imaging. *Optics Letters*, *39*(17), 5150–5153.

Wu, J., Xie, Z., Liu, Z., Liu, W., Zhang, Y., & Liu, S. (2016). Multiple-image encryption based on computational ghost imaging. *Optics Communications*, *359*, 38–43.

Yong, P., & Fu, X.-q. (2016). Ghost diffraction and ghost imaging in two-color ghost imaging. *Proceedings of SPIE*, *10157*, 101570Y.

Yu, H., Lu, R., Han, S., Xie, H., Du, G., Xiao, T., & Zhu, D. (2016). Fourier-transform ghost imaging with hard X rays. *Physical Review Letters*, *117*, 113901.

Zhang, D.-J., Tang, Q., Wu, T.-F., Qiu, H.-C., Xu, D.-Q., Li, H.-G., … Wang, K. (2014). Lensless ghost imaging of a phase object with pseudo-thermal light. *Applied Physics Letters*, *104*(12), 121113.

Zhang, M., Wei, Q., Shen, X., Liu, Y., Liu, H., Cheng, J., & Han, S. (2007). Lensless Fourier-transform ghost imaging with classical incoherent light. *Physical Review A*, *75*, 021803.

Zhang, P., Gong, W., Shen, X., Huang, D., & Han, S. (2009). Improving resolution by the second-order correlation of light fields. *Optics Letters*, *34*(8), 1222–1224.

Zhao, C., Gong, W., Chen, M., Li, E., Wang, H., Xu, W., & Han, S. (2012). Ghost imaging lidar via sparsity constraints. *Applied Physics Letters*, *101*(14), 141123.

Zhu, J., Yang, X., Meng, X., Wang, Y., Yin, Y., Sun, X., & Dong, G. (2018). Optical image encryption scheme with multiple light paths based on compressive ghost imaging. *Journal of Modern Optics*, *65*(3), 306–313.

CHAPTER TWO

The Early Electron Microscopes: Incubation*

John van Gorkom✠, with the cooperation of Dirk van Delft and Ton van Helvoort

Contents

* This article represents a chapter from the draft doctoral thesis of the late John van Gorkom. E-mail for correspondence: hawkes@cemes.fr.
✠ Deceased 2013.

Advances in Imaging and Electron Physics, Volume 208
ISSN 1076-5670
https://doi.org/10.1016/bs.aiep.2018.08.002

> ## 0. INTRODUCTION: THE INCUBATION

After 1933 the German development of the electron microscope came to an apparent standstill. Both Knoll and Ruska had gone into the television industry, von Borries was working for an electricity company in Essen, Brüche had discovered the promising prospects of night vision and Rüdenberg was not allowed by his employer to develop electron microscopes, to name the major players of this story so far. Between 1933 and 1937 only slow progress would be made in the further improvement of the design and performance of the instrument, regardless whether the magnetic transmission type or the electric emission microscope was concerned.

Ruska has called this period of approximately three years "the incubation" (Ruska, 1970). It appears to be a rather appropriate description for this period of stalled development.[1] In reality, it was the silence before the storm. Clouds of new ideas and concepts were gathering, once in a while a flash of progress could be seen, followed by the low rumble of its impact, but for an endless time no revelations would shower down on science.

The incubation was a process that was not confined to Germany, but took place at an international scale. Therefore this chapter is structured along geographical lines, in order not to drown in a sea of events, scattered all over the world. The chronological order of the more crucial moments is applied as a secondary criterion. This means that the chapter starts with Belgium in Section 1, followed by North America in Section 2, the United Kingdom in Section 3, Germany in Section 4, and the Netherlands, Japan and France in Section 5. In Section 6 a summary will be presented, followed by conclusions, which will be related to those of van Gorkom (2018). The USA and Canada are taken together in one section for the reason that the historical developments in these two countries are closely linked.

---

[1] Lin (1995) also adopted the term.

1. BELGIUM

The first and probably most prolonged rumbling was to be heard in Belgium, where the physical chemist Ladislaus Marton had taken a keen interest in the new instrument. Already in 1933, he and Maurice Nuyens had published a general introduction to electron optics in a Dutch journal and had paid ample attention to the transmission microscope (Marton & Nuyens, 1933), as mentioned in van Gorkom (2018). There it was also told that Marton had written a letter to Ernst Ruska on 4 June 1934, in which he stated that he had been intending to contact Ruska for some time, but had wanted to wait for his first articles to appear.[2]

By then, his first articles were indeed on their way. On 7 May 1934, Marton had written a letter to the editor of *Nature*, and on the next day he had presented a paper to the science section of the Belgian Royal Academy, both of which were going to be published. The letter to *Nature* appeared in the edition of 16 June and carried the heading "Electron microscopy of biological objects" (Marton, 1934a). This heading is already of historical significance, since it was the first time ever that such an explicit reference to the biological application of the electron microscope was made.

The short contribution contains "the first histological photographs ever produced by the electron microscope" as Marton himself would write in later years in his "Early History of the Electron Microscope" (Marton, 1968). Two photographs showed cells of *Drosera intermedia*, commonly known as oblong-leaved sundew. The object had a thickness of 15 μm, which is extreme by modern standards, where objects will be 0.1 μm thick at the most. Microscopical sections had been provided by the botanist Marcel Homès and the zoologist Paul Brien, who were friends of Marton at the Brussels Free University (Marton, 1968). The microscope preparations had been fixed with osmium tetroxide (OsO_4). This fixative and dye has become a classic in electron microscopy after Marton's introduction of it. Osmium tetroxide is a heavy metal oxide and therefore very suitable for dyeing in electron microscopy, since it has a strong scattering effect on electrons and therefore it will enhance contrast significantly. In light microscopy, osmium tetroxide has been familiar as a histological dye since the German zoologist Max Schultze applied it for the first time in 1864 (Baker, 1958). Since the osmium tetroxide reacts with the double bonds

---

[2] Ladislaus Marton, letter dated 4 June 1934, personal archive of Ernst Ruska. A copy of it is in my possession.

of unsaturated fatty acids, it specifically dyes the membranes of cells and its organelles, where most of these lipids are found. It was certainly justified, therefore, to speak of the very first histological objects in electron microscopy. At the same time, these first images showed little more than what could be called an osmium skeleton, since all organic matter had been charred or evaporated during a quite lengthy exposure to the electron beam.

The *Drosera* photographs were meant to illustrate Marton's ideas about ways to preserve cellular structure, which he explained in the first two paragraphs of his letter:

> "In a recent paper, Ruska demonstrated experimentally the possibility of surpassing considerably the resolving power of an ordinary microscope by the use of an electron microscope. This high resolving power cannot be applied in biological research, however, without developing a new histological technique to prevent the destruction of the organic cells by the intense electronic bombardment.
> To overcome this difficulty, it seems that there are the following possibilities:
> (1) Intense cooling of the object (for example, by contact with an extremely thin metal foil which is cooled by conduction).
> (2) Impregnating the object with a substance which makes the object less destructible.
> (3) Impregnating the object in such a way that a framework of the object is preserved although the object itself is destroyed.
> (4) Combining methods (1) and (2), or (1) and (3)."

With the osmium impregnation Marton had illustrated possibility 3. As would become clear in later papers, possibility 1 had not been tried yet, while possibility 2 was believed not to be feasible in practice.

The quote also shows that Marton had been fair enough to acknowledge Ernst Ruska's work in the very first sentence. The "recent paper" to which he referred, was the kind of manual on how to build an *Übermikroskop*, which Ruska had submitted on 12 December 1933. At the same time, however, Marton did not mention that in this same manual, Ruska had already made the suggestion to use a metal as a histological dye (Ruska, 1934a). One might wonder, though, how much impression Marton's letter will have made on the average reader. The largest magnification was 450 times, and it merely seems to illustrate that a biological object will be completely charred in the electron microscope, just as any reasonable person would have expected.

On 8 May 1934—one day after writing his letter to Nature—Marton was given the opportunity to present his results to the science section of the Belgian Royal Academy. His contribution was published in the section's journal, called *Bulletin de la Classe des Sciences de l'Académie Royale de Belgique*, and carried the title "La microscopie électronique des objets biologiques", which has exactly the same meaning as the English header of the letter in Nature (Marton, 1934c), making it the second time that such an explicit reference was made to biological applications. The publication date of the *Bulletin* was 15 June, so one day before the publication of Marton's letter in *Nature*. Also in this paper, Marton started by referring to Ruska's surpassing of the resolving power of light microscopy, after which he continued to say:

> *"Our work is undertaken with another objective. As we envisage the application of electron microscopy to biological problems, we set ourselves to the development of a histological technique,—given that the current histological technique is not applicable in our case. As a consequence, when constructing our current microscope, we did not try to obtain very large magnifications, but first of all good optical results."*

This makes it clear once more that right from the beginning it was Marton's very conscious choice to try to develop biological applications.

In this second paper, Marton presented again the four options to preserve cellular structure, which he had already described in *Nature*. This time, however, he had more room to elaborate on them, and explained that option 1—the idea to cool the object by using a metal foil as a support—would be very difficult to realise. Reason for this is the strong scattering of the electrons by the foil, as Ruska had already observed in his *Diplomaufgabe* from 1930, when he likened it to the effect of frosted glass on light falling through (Ruska, 1930). The recognition of the role of scattering would become one of the many smaller priority issues in later years. In his "Early History of the Electron Microscope", Marton claims that his paper in the Belgian *Bulletin* proves that he had been the first to realise the importance of scattering in image formation. He acknowledges Ruska for having made some remarks already about image formation in the "manual" from December 1933 (Ruska, 1934a), but according to Marton, Ruska had not called it scattering then (Marton, 1968). Ruska had used the German word *Diffusion*, which means the same as it does in English. This is certainly not the same as scattering, which is called *Streuung* in German. To someone as distinguished as Dennis Gabor this argument was convincing enough to support Marton in his priority claim, writing in 1968 in the preface to Marton's "Early History":

> *"The electron microscope is (...) entirely different in its action, that is to say in the way it forms contrast in the image. Absorption played almost no part. The main source of contrast was scattering, and the first to recognize this fact clearly was Bill Marton." (Gabor, 1968)*

It is difficult, however, to justify this claim. Already in 1985, it was pointed out by Marton's biographer Charles Süsskind that Ruska had spoken of *Streuwinkel* (scattering angles) in the sentence right before the one in which he had used the word *Diffusion* (Süsskind, 1985). Apart from that, the major importance of these very early views on image formation was the recognition that image contrast did not need to be caused by absorption. As stressed before, the difference between absorption or no absorption is the difference between immediate destruction or no immediate destruction of the object. Whether instead the electrons are scattered or would diffuse was not crucial at that point. It is also questionable whether Marton was really discussing image formation in the *Bulletin*. His remarks on scattering refer solely to the effect of using a metal foil as a support for the object. In the paper, he only intends to explain that an increasing thickness of the foil would result in an increasing loss of sharpness and intensity, which was not due to absorption. He does not spend a word on the way contrast would be rendered in the object itself. Furthermore, if we choose to discuss this issue on the level of semantics, maybe it should also be pointed out that in the *Bulletin*, Marton used the French word *dispersion* to refer to scattering, which is not completely correct. In the fifth and last place, it should be mentioned that Marton made a very contradictory statement in a talk, which he gave four weeks later, on 4 June 1934, saying:

> "Therefore we can make electronic images of any object that emits electrons, or objects which let electrons pass through and thereby make those objects visible by absorption. (Analogy in optics: the image of a lamp, which emits light, or the image of a device which absorbs partially, projected on a screen.)" (Marton, 1934d)

This remark shows no recognition at all of the role of scattering.

This issue about image formation was not the only attempt to challenge Ruska. In the *Bulletin*, Marton finally makes a reference to Ruska's suggestion to use metals to impregnate cells, which the latter had made in the same paper in which he had used the word *Diffusion* (Ruska, 1934a). Marton presents the suggestion as an answer to no. 2 in his list of options to preserve cellular structure (Marton, 1934c), which was an overall conservation of the tissue. Subsequently, he deems it unlikely that this option no. 2 will work (Marton, 1934c). Therefore, he had chosen to pursue option no. 3 on his list, which is to conserve only a skeleton of the cellular structure, allowing the cellular fabric itself to be destroyed. The essence of Ruska's suggestion, however, had been to use a metal as a histological dye, which is essentially the same as Marton was doing with his osmium tetroxide impregnation.

Self-evidently, Marton presented in the Bulletin his images of *Drosera intermedia* cells again, or whatever was left of it after the electronic bombardment, but this time he also elaborated on the design of the microscope that he had used to make the photos. It was the successor of his first primitive device, which he had constructed in late 1932. It consisted of a high voltage generator of approximately 40 kV, and a horizontal brass cathode-ray tube, fitted with three electromagnetic coils, which served as condenser, objective lens and "ocular". The objective lens and ocular were encased in the way as described by Knoll and Ruska in 1932 (Knoll & Ruska, 1932a), although Marton does not make any reference to their paper. The ocular, that is projector lens, was an external coil which could slide along the outside of the tube. This was done to avoid technical complications, as Marton would explain in a later paper (Marton, 1935a). Quite remarkably, he combined this external ocular—not the objective lens—with pole pieces at the inside of the tube. It meant that the actual lens and its pole pieces were separated by the wall of the tube, implying that part of the magnetic field would leak away through the space that was taken up by the brass wall. As a matter of fact, this construction had already been suggested by Knoll and Ruska in their last mutual paper, as an alternative in case it was necessary to use an external coil (Knoll & Ruska, 1932b). Nevertheless, in later years Marton raised the suggestion, that he had come up with this idea independently (Marton, 1968; see also Süsskind, 1985), although Knoll and Ruska's last paper was already cited by Marton and Nuyens in 1933 (Marton & Nuyens, 1933).

Despite all of this, again we should acknowledge Marton's enthusiasm and untempered belief in the future of biological electron microscopy. The last page of his paper in the *Bulletin* carried the heading "Perspective" (Marton, 1934c). Here Marton wrote about the new instrument:

"Thanks to its extremely high resolving power, it will be possible to make great discoveries, not only in biology and bacteriology, but also in most domains of the other natural sciences."

And:

"Especially the application of this instrument in the biological domain makes a close cooperation between physicists and biologists desirable."

It is the first time that bacteriology is mentioned in connection to the electron microscope. It is also absolutely clear that Marton and Ruska shared the same vision.

These two publications on "the electron microscopy of biological objects" were only the first of many more to come. As already indicated above,

Marton presented another lecture on 4 June 1934, on the same day that he had written to Ruska. It was a talk for the Royal Society of Medical and Natural Sciences of Brussels and published in their "Annals and Bulletin" (Marton, 1934d). It carried the telling title "The electron microscope—first attempts to apply it to biology", and it is beyond doubt that Marton was entitled to call it this way. The paper started with a long explanation of the differences between light and electron microscopy. In this respect, it was another general introduction to electron microscopy, just like Marton's first paper with Nuyens (Marton & Nuyens, 1933), the simultaneous paper of Zworykin in the *Journal of the Franklin Institute* (Zworykin, 1933), and Ruska's short article in *Forschungen und Fortschritte* (Ruska, 1934d), which had all been written in the previous year. The introduction was followed by a description of the instrument and the same list of four options to preserve cellular structure, which had already been presented twice before. Once more, Marton presented the images of *Drosera intermedia* which had appeared in *Nature* and the *Bulletin*, but he also had a new image to offer. Finally, he had been able to get hold of aluminium foil that was only 0.5 μm thick, enabling him to try option no. 4, that is, imaging osmium impregnated cells on a cooled foil. This time he had used cells from *Neottia nidus-avis*—commonly known as bird's-nest orchid—and managed to produce images of cells which also showed nuclei for the first time. This progress can be explained by the fact that the interior of the cells was supported by the foil now, so it would not fall through immediately upon exposure to the electron beam. A price was paid as well, since the foil would cause a significant blur of the image.

In this paper, Marton also made the first announcement of what would become an electronic shutter. Straight from the beginning, he had continuously encountered the problem that he needed some time to focus the instrument on the object, which actually meant that most of the object had already been destroyed by the time he was ready to make a photograph. For his osmium impregnated cells this mattered less, since there was still the osmium skeleton to see, but for a more sophisticated histological technique, he would need far shorter exposure times. Solution would be, he thought, to use a dummy object to focus on, than to deflect the electron beam, replace the dummy by the real object, and switch the beam back for a controlled period of time. For this he needed a kind of deflector as well as an object chamber with which he could easily exchange the objects. Simply switching the beam on and off was no option, since it needed too much time to stabilise.

Meanwhile, Marton's references to any of his German colleagues were becoming very scarce. His bibliography only mentions two of his own papers and a book by Ernst Brüche and Otto Scherzer which had been published in early 1934. In the text, Ruska is mentioned once for having made 12,000 fold magnifications of "*in*organic objects" (Marton, 1934d).[3] This is a peculiar remark, since Ruska makes it very clear that he had used cotton fibre for his experiment.

Also this lecture for the Brussels Royal Society was concluded with the hopeful expectation that the electron microscope would help to unravel the mysteries of "the very small" in the fields of biology, bacteriology and other natural sciences, especially if physicists would work together with the specialists who could apply the instrument. However, his optimistic view was not shared by everyone in his audience, as Marton recounted in his "Early History":

> *"After I presented what I believed to be a good case for a new tool, the eminent bacteriologist and Nobel Prize winner Jules Bordet (1870–1961) rose and said: 'Oh no, no! Let us not have an electron microscope—it's troublesome enough to interpret the images we get by the light microscope.'"* (Marton, 1968; see also Marton, 1976)

It was going to be the first of many more critical responses from the intended target group.

It must have been immediately after this lecture that Marton left for Hungary with his wife Claire, as he had written to Ruska that same day. The planned visit afterwards to Berlin is also described in Marton's "Early History":

> *"Towards the end of June of 1934 my wife and I visited Berlin and became personally acquainted with Knoll, Ruska, Brüche, and several others active in electron optics. (. . .) When I showed my micrographs of biological objects it was very gratifying to see initial unbelief ("This is not possible!") give way gradually to requests for prints.*
> *I have looked up the dates of this trip to Berlin, because of some of the circumstances surrounding it. The atmosphere in Germany was quite electric. One of our friends (not connected with electron microscopy) took us miles out of Berlin in his car to a place where one could see at least a mile in each direction, stopped there, closed all windows, and said: 'Now we can talk.' Small wonder, the date was 30 June 1934, the 'Night of the long knives'—the great blood bath of Nazi Germany, when Himmler's SS massacred the leadership of Röhm's rival SA."* (Marton, 1968)

The quote not only illustrates the immediate contact between Marton and his German colleagues, but also shows his awareness of the political situation. This might well explain why he would decide to leave Europe several

---

[3] Ruska (1934a) published three different photos of cotton fibres, all carrying the caption *Baumwollgespinst*.

years later. Three days after the Night of the Long Knives, the *Gesetz über Massnahmen der Staatsnotwehr* (Law on Self-Defence Measures by the State) came into power in Germany.[4] It contained only one article:

> "As self-defence by the State, the measures that were taken to suppress the attacks of high and national treason on 30 June, 1 and 2 July 1934, were lawful."

From then on Hitler was free to do anything he liked for reasons of "self-defence". Halfway 1934 already, he had become absolute dictator.

On 7 August 1934, Marton submitted his next publication. This time it was a letter to the editor in the 15 September issue of *Physical Review* (Marton, 1934b), the same American journal in which Calbick and Davisson had published their first abstract on electron lenses in 1931. It was Marton's fourth article of 1934. All four of them had been dedicated to the biological application of electron microscopy, while three of them even carried exactly the same title, "Electron microscopy of biological objects." Repetition is a very effective way to get your message across and Marton seems to have been aware of it. He started his letter by giving for the fourth time his four options to preserve cellular structure, while exposing it to the electron beam. Subsequently, he presented his results. Two pictures of *Neottia nidus-avis* cells were shown, although the cells were called "seaweed" by the journal instead of bird's-nest orchid for some obscure reason.[5] One was a 1000-fold magnification of the "skeleton" of cells impregnated with osmium. The other one, showing a magnification of 500 times of impregnated cells on cooled 0.5 μm thin aluminium foil, is reproduced in Fig. 1. Both images had not been published before. This time, Marton announced that he was going to build a new microscope, which would allow him to reduce the exposure time. The significance of this letter in the highly esteemed *Physical Review* is certainly the fact that Marton introduced the results of biological transmission electron microscopy to an American audience for the first time. His short letter contains only one reference, and that is Marton's first publication in Nature, only three months before.[6]

On 11 November 1934 Ladislaus Marton wrote to Ruska again, apparently answering a request for photographs. In German he said:

---

[4] Gesetz über Massnahmen der Staatsnotwehr, *Reichsgesetzblatt*, Teil 1 (1934) 529.

[5] The fact that it is *Neottia nidus-avis* can be learned from Marton (1935a), as well as from Marton (1935c).

[6] The reference contains a typing error, which had changed Marton into Martin. It is a bit ironical, since one year later England would give rise to its own first pioneer with that name.

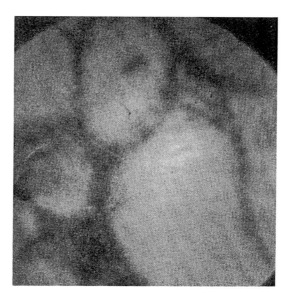

Figure 1 Image showing bird's-nest orchid cells on aluminium foil. From Ladislaus Marton, Electron microscopy of biological objects, *Physical Review* 46 (1934) 527–528.

"Dear Herr Dr. Ruska, excuse me for answering your letter only just now, but I had hoped to send you the desired photographs simultaneously. As my new apparatus is still not finished yet, unfortunately I can only send you older photographs, which you know for some part already. I hardly believe that I will have interesting results before the end of November, but if I might have something, I will not omit to send you copies right away. Now I also want to make a request to you. As I have to give two lectures in France halfway December, I would be very pleased, if I could take your results there as illustrations. Could you please send me, on my account of course, some slides (let's say three or four) of things that you consider very interesting. Of course I will not fail to mention the source." [7]

First of all, it is another illustration of Marton's direct acquaintance with Ruska, but it is also interesting to read that Marton was going to give talks in France. One of these talks was given for the French Physical Society and was already published by the end of the next month (Marton, 1934e). This implies that by the end of 1934 Marton had been spreading the word to the Netherlands, Great Britain, Belgium, the USA and France. Apart from this, the letter gives a second reference to the new microscope he planned

---

[7] Personal archive of Ernst Ruska.

to construct, which would become his third device in about three years time.

After his talk for the French Physical Society, Marton was invited to write a review for the French journal *Revue d'Optique*, which was published in April 1935, named "The electron microscope and its applications" (Marton, 1935c). It is another general introduction, followed by details of his own work again. This time, Marton was finally able to present photographs of new parts of his next microscope. These parts were an object chamber with airlock, an airlock for inserting photographic plates into the vacuum, and a beam deflector, which was to be used as an electronic shutter. The deflector had already been announced, as we have seen above, and probably this is the only novelty in instrumental design, for which Marton deserves the credit of having been the first to come up with it. Simultaneous with Marton's presentation in *Revue d'Optique*, von Borries and Ruska had published a review in Germany in which they too discuss a deflector with the same purpose (von Borries & Ruska, 1935), but Marton had already described the idea in June 1934 (Marton, 1934d) and actually built one in the mean time. The idea of airlocks for objects and photographic plates, however, had just been patented by Ruska a few months before, as will also be discussed in Section 4 (Ruska, 1934b, 1934c), while the need for such airlocks had already been expressed by Ruska in his 'manual' from December 1933 (Ruska, 1934a).

From this review, it also becomes clear that Marton had not been able to do much work on histological techniques. The article contains five images of cells that had all been used before, including the two images of bird's-nest orchid that had been published in the *Physical Review* and an image of sundew that had even appeared three times before: in *Nature* and in both Belgian articles from 1934. Probably the most important quality of this article by Marton is the fact that it is the first major discourse on electron microscopy that was published in France.

In his "Early History" from 1968, Marton has given an explanation for his rather slow progress at the time:

"(...) Each bit of progress cost a tremendous amount of effort. University funds were practically nonexistent and with the meager resources I had (thanks to family support) Mrs. Marton and myself became frequent visitors at the flea market. A reasonable amount of German equipment was available that had been abandoned after the war of 1914–1918 and numerous components of the new microscope benefited from this availability at very low prices. The only other source of money was a modest grant from the Institut International de Physique Solvay. Professor Henriot was at that time permanent scientific secretary at the Solvay Congresses in

Physics and thanks to his intervention I was for some years the recipient of a Solvay fellowship." (Marton, 1968)

Altogether, since his last results in the *Physical Review* of September 1934, it was going to take Marton nearly two years before he would be able to present histological results again. This did not mean that he ceased to publish in the mean time. In Belgium two more articles on electron microscopy appeared halfway 1935. Both of them turned out to be part of a sequel that had started with Marton's first Belgian article in the Bulletin of the Belgian Royal Academy. The two new ones were therefore titled "Electron microscopy of biological objects (Second communication)" and likewise "(Third communication)".

Part II had been presented by Marton at a session of the Academy on 7 May 1935 (Marton, 1935a), one year after the first part. It was a report on the limitations of his way of practising electron microscopy: the limited magnifying power of his microscope, the limits to its resolving power and limits to the irradiation of his objects. When discussing the magnifying power, he finally acknowledged that the combination of an external projector lens with internal pole pieces was not ideal, since this will give a larger focal distance. As far as the resolving power of his instrument was concerned, he estimated it to be 0.6 μm, which is close to the theoretical limit of a light microscope, but certainly not beyond it. And with respect to the irradiation of his objects, he mentioned for the first time that he had started to use a more sensitive viewing screen and more sensitive photographic material to capture the image on the screen. This had resulted in a 50-fold reduction of the exposure time, which probably means that the exposure time had gone down by 98 percent. Only in his conclusion, Marton briefly mentioned the electronic shutter he was working on.

A more curious detail in Marton's paper is his report on an attempt to image living objects in the electron microscope (Marton, 1935a). It seems to have been primarily based on the widespread misconception that live objects are essential to biological microscopy. Certainly, there are enough reasons to put living objects under a microscope, but at the same time the majority of biological material studied under the light microscope is thoroughly fixed, and therefore absolutely dead. Marton was not the only physicist who thought that it was vital to be able to study living matter, but he was the first who actually tried to do it. The idea was to use Lenard windows for it, the action of which has already been discussed in van Gorkom (2018). From a material point of view, a Lenard window is nothing else but a very thin metal foil, such as Marton had already been using as a support for his objects. So basically, his experiment implied that he only had to put a second foil on top of the object, which at least would prevent the object from being dehydrated, he believed. Marton was fair enough to report that his attempts failed completely—he could not see a thing.

As a matter of fact, the idea to use Lenard windows for this purpose had not been that unique. In the above-mentioned German review that von Borries and Ruska had published in April 1935, they also discussed this same idea, although they immediately added that the foils should not be thicker than 0.01 µm, which is 50 times thinner than Marton's foils (von Borries & Ruska, 1935). However, it was not von Borries and Ruska's idea, as Reinhold Rüdenberg had described the concept for the first time in one of his Siemens patents from 1932 (Rüdenberg, 1932).[8] Von Borries and Ruska made the effort to acknowledge this in a footnote.

Part III of Marton's series was presented by him at a session of the Academy on 1 June 1935 (Marton, 1935b). Finally he was able to present his microscope no. 3, according to his own numbering; being his second two-stage instrument and the first vertical one. The improvements were considerable, as it appears. Marton had incorporated a more stable electron gun, which was powered by a 90 kV generator, as can be learned from a later paper (Marton, 1937). He was using internal lenses now, with pole pieces for both the objective lens and the projector lens. His deflector with electronic timer was described in detail now, and also all details of his new object chamber were given. As he stated himself, this object chamber had started with the idea of the revolving object holder in Ruska's 'manual' from December 1933. Subsequently he had expanded the concept with the idea of an airlock, in this way ending up with an airlock for a revolving object holder. The result was, of course, a rather complicated device, and you may wonder whether it was actually necessary to combine these two concepts. If you have an airlock, making it possible to replace objects, you do not really need the revolver anymore. Just to compare: Ruska first came up with the revolver, and then replaced the concept by an airlock for a single object, as we will see in Section 4. Marton concluded his paper with a detailed description of a second airlock, which enabled internal photography. Altogether, he was able to present in June 1935 the most advanced transmission electron microscope of its time, even if most of the innovations had not been his own idea. Unfortunately, for the article's reader it will not

---

[8] Von Borries and Ruska referred to the Austrian Siemens-Rüdenberg patent 137 611, which had already been published in May 1934

have been easy again to tell where all ideas came from. Given the fact that the complete layout of the machine—with the exception of the electronic shutter—had already been discussed in Ruska's 'manual', you would expect more credit for the latter, but Marton's only reference to Ruska concerned, as said, the revolving object holder. And while the lenses are entirely based on the published designs by Knoll and Ruska (1932a, 1932b), and Ruska alone (Ruska, 1934a), no source is given at all. All other references in the paper relate to engineering issues of a more general nature, like electron guns or internal photography.

In June 1935 Marton had still not been able to show new results, and it would take another year before this was going to happen in the May–June 1936 issue of *Revue de microbiologie appliquée à l'agriculture, à hygiène et à l'industrie*—a journal published for a few years by the French Ministry of Public Health. The article was simply called "The electron microscope" (Marton, 1936b) and started as yet another general discourse on the underlying principles of the instrument and its construction, which was illustrated with images of Marton's new microscope. New as well was the comparison of the same type of tissue imaged by a light microscope and by an electron microscope. This had been done before by Ruska in 1931 for a metal grid, and it was also done by others for cathode surfaces, but not for biological objects yet. Marton presented a recent electron image of *Neottia* cells, together with a light microscopical image of the same cell type. They were not the same cells in both images, as it was practically impossible at the time to make a light image of specific cells in a specimen and then try to find back exactly the same cells again, meanwhile exposing them to the electron beam. Making this kind of comparisons, however, was a sensible thing to do, as it is certainly the most convincing way to show that electron microscopical images are reliable. The comparison was successful to the extent that the electron image looks rather similar to the light image.

Another new element was the absence of osmium tetroxide. Marton did not mention this in the text, and it would only become apparent in another article, published half a year later, in which he published the same images (Marton, 1936a). From that same source, it also becomes clear that he had introduced a new type of support for his objects. He had used zapon varnish, which is a kind of celluloid—a solution of nitrocellulose in amyl acetate. It was a technique that originally had been developed for infrared windows, as Marton explains in his "Early History" (Marton, 1968). When you drip zapon on the surface of water, the nitrocellulose turns into a very thin film of varnish. Such a film proved to be a far better support for objects, and in later years this method would become a standard method worldwide. The afore-mentioned new electron microscopical image of *Neottia* that Marton published, shows a little crack, which happened to be a fortunate incident, since it allowed him to make an estimate of the resolution of his microscope. He believed it to be 0.02 μm, which is approximately ten times better than can be achieved with a light microscope. It is also in this paper that Marton is using the word French word *diffusion* (scattering) for the first time, while discussing image formation. In the conclusion of his paper, once more he discussed the application of the electron microscope to bacteriology. It would take some more time, however, before he was actually going to try this himself. Again, the paper contains no reference at all to the work by Ruska and colleagues, despite the fact that it is first and for all a general introduction on the transmission electron microscope.

Nearly three months later, Marton wrote a third letter to Ruska.[9] In the letter, dated 29 August 1936, Marton gave a short report of the latest developments. He writes that he had started to use unstained cells, leaving out the osmium, and he had been able to bring the exposure time of his objects down to values between 0.1 and 0.003 seconds. These short exposures allowed him to study untreated cells, like red blood cells, which had not been fixed at all. He refrained from sending Ruska the images of these latest attempts, since they were too bad, because of his lack of adequate biological skills, Marton explained. Instead he sent some images, which he believed to be more representative. It appears, the letter was an answer to a recent request by Ruska, who was going to present a lecture two weeks later at the 12th German Conference of Physicists and Mathematicians in Bad Salzbrunn. At that occasion Ruska would show Marton's cracked Neottia image together with the light image for comparison (Ruska, 1937).

A week later, on 5 September 1936, Marton submitted his next paper, which appeared in the November 1936 issue of *Physica*, a Dutch journal. It was in French again as *Physica* used to publish in several languages. This time it was a purely theoretical treatise of "Some considerations concerning the resolving power in electron microscopy" (Marton, 1936c). Marton had become fully absorbed by the theory of image formation now and presumed that a true electron image is formed by scattering alone. The paper was an attempt to give calculations for the effect of the object's thickness in relation to electron speed, relative density of the object and its average

[9] Personal archive of Ernst Ruska.

atomic weight and atomic number. His major outcome was that the higher the electron speed is, the poorer the scattering. In other words, the faster an electron is, the less possibility there is to deflect it. And the poorer the scattering is, the poorer the contrast in the image. He therefore concluded that electron acceleration had opposite effects. The faster electrons are, the shorter the wavelength is, and the better the resolving power will be in terms of Abbe's formula. But at the same time, the faster electrons are, the poorer the contrast gets. The positive effect of speed he called "surface resolving power", which is the resolving power we normally refer to. The counter effect, he coined "depth resolving power." On the fly and out of the blue he also mentioned the idea of using a film of zapon on which to place an object, as part of his theoretical exercises. This made it the first time that this idea appeared in print.

Exactly three months later, on 5 December 1936, Marton presented number 4 of his series of talks for the Belgian Royal Academy, all called "Electron microscopy of biological objects" (Marton, 1936a). It started with a rather annoyed introduction, announcing that he was going to explain some fundamentals, which he had assumed to be too obvious to treat in previous papers. This remark is followed by a long discourse on the focal distance of an electron lens and its relation to the strength of the magnetic field and electron speed. Subsequently, the paper gives more details about his new images of *Neottia* cells and the use of zapon films as a support. The zapon films were between 0.1 and 0.2 μm thick, he reports, and certainly less dense than the aluminium foil of half a micrometre thick that he had tried in the past. It is also mentioned now that the cells were not stained with osmium anymore. His estimation of the smallest visible detail is 0.01 μm, about twenty times smaller than a light microscope would show (Marton, 1936a). This is a bit surprising, as he had given a value of 0.02 μm half a year before, in the French article in which he had shown one of the new images for the first time. In the references at the end of his article, he referred this time to a paper by Friedrich Krause, which will be discussed in more detail in Section 4. This paper of Krause had appeared only shortly before in the *Zeitschrift für Physik* (Krause, 1936), meaning that Marton was well aware of recent developments in Germany at the time.

Marton's fifth and final part of his series "The electron microscopy of biological objects" was presented to the Belgian Royal Academy on 5 June 1937 (Marton, 1937). The paper was very short and mainly dedicated to the presentation of images of *Chromobacterium prodigium*, blood red bacteria, nowadays called *Serratia marcescens* (Purkayastha & Williams, 1960). The magnifications were only 750 times and showed little more than specks, which lay on a nitrocellulose film of 0.1 μm thick. Nevertheless, Marton can be credited for having been the first in history to make electron microscopical images of bacteria.

The paper turned out to be Marton's worthy goodbye to Europe and its growing intolerance. In the autumn of 1938, Ladislaus Marton and his wife Claire emigrated to the USA. It was the conclusion of pioneering years, in which Marton had played a crucial role as very early ambassador of biological electron microscopy.

2. NORTH AMERICA

At the start of 1934 not much had happened yet in North America, as far as geometrical electron optics was concerned. As John Reisner puts it in his "Early History of the Electron Microscope": "The German successes did not stimulate much activity in the United States scientific community" (Reisner, 1989). And neither did they in Canada. In 1931 a short abstract by Davisson and Calbick had been published in *Physical Review*, in which they concluded that an electrostatic electron lens is the analogue of a spherical lens in light optics (Davisson & Calbick, 1931; a year later an adjustment was published in Davisson & Calbick, 1932). And in van Gorkom (2018) we also saw that in 1933 an article called "On electron optics" was published by Vladimir Zworykin, in which he discussed cathode emissions and the focussing properties of electrostatic lenses. After 1933, however, electron microscopy slowly started to gain substantial momentum.

2.1 Canada: Wilfrid Benham

The first follow-up was a presentation by Wilfrid E. Benham on 7 March 1934 for the Wireless Section of the American Institution of Electrical Engineers. The paper, called "Note on a demonstration of a low-voltage electron microscope using electrostatic focusing", was submitted to the Journal of the Institution on 28 March and in its final form resubmitted on 30 June 1934 (Benham, 1934). As it appears, Benham had built an emission electron microscope for the Canadian branch of the Marconi Company,[10] inspired by Zworykin's observation that "any form of high-vacuum cathode-ray oscillograph can generally be made to give an enlarged

---

[10] In the paper, it is mentioned that permission for publication had been obtained from Marconi's Wireless Telegraph Co. Ltd. From the Canadian Intellectual Property Office

image of the cathode", as Benham phrases it in his introduction. His other source of inspiration was the report by Davisson and Calbick. The latter had defined focal distance f as $4V/(G_2 - G_1)$ for a hole in a charged plate, where V is the kinetic energy of the electrons and $G_2 - G_1$ "the difference between the potential gradients on the emergence and incidence sides of the plate", as the authors defined it (Davisson & Calbick, 1932). On the basis of this formula, Benham explained that an electron lens could be convex as well as concave, since the formula does not prevent the focal distance from becoming negative. Apparently he was not aware of the observation by Johannson and Scherzer a year before that such a concave lens cannot exist, despite the fact that the paper in which Johannson and Scherzer prove this is listed in Benham's bibliography. Benham's microscope operated at a low voltage of less than a kilovolt with which he obtained some enlarged images of the cathode. His results led him to the conclusion that "Davisson's formula does not appear to give good quantitative agreement with experiment." His view on the use of the electron microscope is just as plain: "The major application of the electron microscope is to the study of thermionically active materials." Considering the fact that he represented an electrotechnical company, this statement is not really a surprise.

2.2 USA: Davisson and Calbick

At approximately the same moment, Chester Calbick and Clinton Davisson themselves presented new results at the Washington Meeting of the American Physical Society which took place from 26 till 28 April 1934. Apparently inspired by Brüche and colleagues, they had used their electrostatic lens to build a two-stage electron microscope for imaging cathode emission. An abstract of their presentation was published shortly afterwards in the 15 May issue of *Physical Review* (Calbick & Davisson, 1934), the same journal in which they had made their previous reports, and in which Ladislaus Marton was going to announce his work on biological application four months later. It is tempting to explain Davisson and Calbick's specific focus on emission microscopy from the fact that they too were employed by industry—in their case Bell Telephone Laboratories, Inc.

it can be learned that Wilfrid Earnshaw Benham was the inventor for Canadian patents 354793, 362580 and 384339, which are owned by the Canadian Marconi Company (http://brevets-patents.ic.gc.ca/opic-cipo/cpd/eng/introduction.html, accessed 30 June 2013).

2.3 USA: Anderson and Fitzsimmons

Going by the thesis of two electron microscopical schools which I put forward in the conclusion of van Gorkom (2018), a focus on biological transmission microscopy should only be found at the universities of North America. The first evidence that this is indeed the case can be found in Reisner's "Early History" (Reisner, 1989). We already saw in van Gorkom (2018) that Paul A. Anderson at Washington State University had taken a keen interest in Ruska's work on the electron microscope, which had inspired Anderson's co-worker Kenneth Fitzsimmons to start build one of their own. Progress was slow, however, Reisner tells, and the instrument was only completed by the end of 1935, although it seems that the project must have started soon after Anderson's appointment as professor and chairman of the Physics Department in 1931 or 1932.

It was a very sophisticated instrument for its time, according to Reisner, with vertical column, three magnetic lenses, a condenser included, extra coils for beam alignment, internal photography, operating at 30 kV and a resolving power equivalent to a light microscope. The first recorded photographs were made in December 1937. Reisner is very specific about the goal of this so-called Fitzsimmons–Anderson microscope:

> "(. . .) from the beginning the primary end in building the microscope was to acquire an instrument that could do microscopy of biological materials. Building the microscope was merely a means to that end." (Reisner, 1989)

Unfortunately they never reached this goal. Somewhere mid–1938 they abandoned the project, as Anderson wrote to Reisner in 1987:

> "By the time we were getting low magnification imaging, other groups had moved well ahead of us, and their lead was increasing rapidly. (. . .) we agreed there was no point in continuing our work on it." (Reisner, 1989)

Fitzsimmons and Anderson never published anything about their attempt to build an electron microscope. Only 25 years later were the remains of the instrument rediscovered in the attic of the Physics Department.

2.4 USA: Scott, McMillen and Packer

Fitzsimmons and Anderson's attempts are not the only example of a university that wanted to build its own electron microscope for biological purposes. A second one is the Washington University of St Louis, where one should not confuse this Washington University in St Louis with Anderson's Washington State University. The initiative in St Louis was led by Gordon H. Scott, who was a professor at the Department of Pathology. According to Reisner, Scott:

"(...) was interested in the role of trace elements in muscle, nerve and other tissues and felt that the electron microscope might help give chemical identification, on a microscopic scale, of the metallic elements whose location was revealed by microincineration." (Reisner, 1989)

With post–doctoral fellow J. Howard McMillen, he started in 1935 to construct a magnetic emission microscope with one lens. A report of their results, called "A magnetic electron microscope of simple design," was submitted on 5 April 1937 to the *Review of Scientific Instruments* (McMillen & Scott, 1937). It starts with the rather unfortunate remark:

"Since the appearance, in 1927, of the first magnetic electron microscope by Busch, many improvements have been made in the details of the original design."

The instrument they built could be described as a hybrid of Brüche's electrostatic emission microscope and Ruska's magnetic transmission microscope. At the one hand, it was fitted with a magnetic lens of the design as described by Knoll and Ruska in 1932, and the instrument was clearly intended to serve a biological purpose, as they remark in their introduction:

"In connection with an investigation of ashed sections of biological materials it was necessary for us to set up an electron microscope arrangement in which only a relatively small magnifying power was required and in which the image was to be formed by the thermionic emission from surfaces whose currents were very small."

At the other hand, it was a true emission microscope, as is illustrated by this same quote, which worked at a low acceleration voltage of only 1 to 2 kV. The biological material was mounted directly on the cathode, where it would be incinerated, after which certain minerals would start to emit. The charring would not affect the distribution pattern of metallic elements, Scott believed. The idea of mounting the object directly on the cathode itself, reminds us of the Knoll, Houtermans and Schulze patent from 1932 on the photoelectric electron microscope, in which material is mounted on the cathode as well (Knoll, Houtermans, & Schulze, 1932c, 1934), and of Knoll and Ruska's third joint article from 1932, in which they refer to the same idea (Knoll & Ruska, 1932b). Notwithstanding these resemblances, it is clear that Scott was trying to develop a rather unique kind of biological electron microscopy.

Howard McMillen left the project in 1937 and was succeeded by Donald Packer, who experimented with a second magnetic lens, and separated the object from the cathode, herewith transforming the instrument from an emission microscope into a magnetic transmission microscope, while still having the study of biological objects in mind. For this reason it seems reasonably fair to label Packer's superior Scott as another representative of biological magnetic transmission microscopy. Altogether, Packer was not very successful and left the job after a year. His successor was Sterling P. Newberry (Reisner, 1989).

2.5 Canada: Burton and Hall

If talking in terms of the gathering of clouds of scientific insights during the incubation period between 1933 and 1937, it appears that by far the most significant developments were to take place in the far north of the American continent.

At the University of Toronto, professor Eli Franklin Burton was head of the McLennan Laboratory, which was part of the Physics Department. Burton had graduated in mathematics and physics from the University of Toronto in 1901 and from 1904 till 1906 he had spent two years with J.J. Thomson—the discoverer of the electron—at the Cavendish Laboratory in Cambridge. This might well explain, of course, where Burton's interest in electron optics originated from. According to John Watson—one of his students—Burton was a friend of Walter H. Kohl, a German doctor in engineering physics, who was employed in Toronto by the electrotechnical company Rogers Radio Tubes Ltd. At Burton's invitation, Kohl gave guest lectures in electronics at the physics department, and on 9 April 1934 already, he had demonstrated a self-made emission microscope there (Burton, 1948; Watson, 2013; Hall, 1985).

One of the other students present at this demonstration had been Cecil E. Hall, just like John Watson in later years a well-known pioneer of American electron microscopy. Hall noted in a personal recollection, published in 1985, about his first acquaintance with the subject:

"The opinion of other graduate students was largely negative. They believed that electron optics was too ill-defined, although it might have some application in the future 'if television ever became practical.' This fitted my thinking too, except my outlook was more optimistic. I think the main aspect that frightened them away was the fact that Professor Burton voiced the strange conviction that an electron microscope could be built that could be used in biological and medical research." (Hall, 1985)

It illustrates that Burton belonged to the same small party of believers in true transmission electron microscopy of which Rüdenberg, Ruska, von Borries and Marton were the other members. According to John Watson, Burton had even attended a conference in Berlin in 1935 to learn more about the application of electron microscopes (Watson, 2013).

Despite his initial doubts about the practical value of electron optics, Cecil Hall was interested to give the imaging of colloids a try as it was fairly easy to realise in terms of time and resources, and most importantly in terms of results. As Hall writes:

> *"Results in the study of colloidal particles could be guaranteed. Biological substances was a different matter. If an application in medicine could be demonstrated even in a small way, we could reach a source of financial support in which our financial needs would be counted as trivial. First, we had to build a microscope and show results." (Hall, 1985)*

It appears, the idea must have come from Burton, as he had earned an additional bachelor degree for work on colloids during his stay at the Cavendish.

To get some practice, Hall first built an emission microscope, based on Helmut Johannson's description of the immersion lens (Johannson, 1933). This was in 1935, so it must more or less have coincided with Burton's trip to Berlin. A year later, time had come to take the next step:

> *"Our goal had to be a high-resolution transmission electron microscope. We knew from the work of Knoll and Ruska (18) and Ruska (22) that such an instrument could be built, but we did not know whether it could be used for serious work in biology. (...) In September, 1936, I decided to put together the simplest possible magnetic microscope for the sole purpose of experimenting with specimen technique and getting some pictures of biological material. I predicted that a resolution of 200 Å (2 × 10⁻⁶ cm) should be easy to achieve—and that is ten times better than the best light microscope." (Hall, 1985)*

Reference 18 is the third Knoll and Ruska article, which gave the first estimate of the resolving power and the first description of a pole piece lens. Reference 22 is Ruska's 1933 so-called manual (Knoll & Ruska, 1932b; Ruska, 1934a). Burton's original plan to image colloids had apparently been abandoned in the meantime, maybe because of new developments, like the work of Marton, and recent German work, which will be discussed in Section 4. Hall's efforts resulted in a horizontal machine with external magnetic lenses and a very modest 5 kV accelerating voltage, which could render magnifications of about 5000 times. According to Hall, its resolving power lay in the range of 200 to 400 Å, which is indeed ten times better than a light microscope, but it turned out to be impossible to make proper transmission images of specimens, which will have been due to the low voltage.

Meanwhile Burton had not managed to raise the necessary funds to keep the project going, and shortly afterwards, Hall was offered a position on the research staff of Kodak Laboratories in Rochester, New York, as this company was considering the idea of building an electron microscope of their own. After Hall started in his new job on 1 February 1937, it would take nearly a year before new students would be able to take on Burton's electron microscope project in Toronto again.

2.6 USA: Vladimir K. Zworykin

Quite remarkably we have to go to Berlin again to learn more about Vladimir Zworykin's interest for electron microscopes. It turns out that Zworykin had come over to Berlin from Camden, New Jersey, to present a lecture for the joint session of the *Deutsche Gesellschaft für technische Physik* (German Society of Engineering Physics) and the *Physikalische Gesellschaft zu Berlin* (Berlin Physical Society) in the "New Physics Auditorium" of the Technische Hochschule on 26 February 1936. The talk was published in the June issue of *Zeitschrift für technische Physik* and was called "Electron optical systems and their application" (Zworykin, 1936).

The article begins with a short report on the introduction that had been given by Karl Mey, the President of the *Gesellschaft für technische Physik*. Zworykin is presented by him as RCA's famous inventor of the iconoscope, which was the name for his television camera design. Apparently, also in science the atmosphere was already politically charged in early 1936. Mey told his German audience that, considering "the current moment", the session would not be the place to compare latest scientific developments in Germany and the USA. As he expected that Zworykin was not going to report on the latest American insights, he added it would be good not to tell him too much about German developments. Therefore, the audience was requested to only ask questions with immediate connection to the lecture. From Mey's remarks it is also clear that Zworykin did not speak German. One may wonder whether he received a full translation of Mey's introduction.

Zworykin started his talk with ranking some of the pioneers of electron optics. He placed the German Hans Busch and the Americans Clinton Davisson and Chester Calbick first, followed by Max Knoll and Ernst Ruska at the Technische Hochschule, and in third position AEG's Ernst Brüche, Helmut Johannson and Otto Scherzer. Subsequently, he reproduced Davisson and Calbick's formula for the focal distance of an electron lens, and just as Benham had done before, he concluded from this formula that the focal distance could be negative as well. Members of the German audience will have realised that this idea was false, but maybe Zworykin told this on purpose, if we give any weight to Mey's introductory warning.

According to Zworykin, two applications of electron optics were the most prominent. These were the "electron beamer" and the electron microscope. He used the term electron beamer as denomination for those cathode-ray tubes that operate with a concentrated electron beam, of which the oscilloscope and television tube are the most obvious examples. His idea of an electron microscope had evolved into the broader concept of a device to make substantial enlargements of small objects, in contrast to his limitation to emission microscopy three years earlier. Noteworthy is the schematic drawing of an electron microscope that he used as an illustration and which he had found in the book *Geometrische Elektronenoptik*, a comprehensive review, published in 1934 by Ernst Brüche and Otto Scherzer. The image shows a microscope design by Walter Knecht from the AEG Research Institute (Brüche & Scherzer, 1934; see also Knecht, 1934), who had built an emission microscope which was fitted with electrostatic lenses, as well as magnetic and glass lenses in order to compare the different imaging systems.

Zworykin's major interest was nevertheless still focussed on electrostatic imaging, and therefore he did not elaborate on the electron microscope as such any further, but instead liked to discuss the efforts at RCA to render distortion-free electron images, followed by a long discourse on electronic multiplier tubes—the latter being a topic that goes far beyond the scope of this article. The details of the attempts to improve the quality of electron images are quite fascinating, however. The focus lay on photoelectric imaging, and one tested application was an electron telescope, illustrating perhaps that the earliest ideas of Rüdenberg, Knoll and Brüche about electron optical applications had not been that farfetched after all. Zworykin's telescope would capture a light optical image with a photosensitive cathode. The latter would translate the image into an electron image, which could be magnified to some extent. From here it was only a small step, of course, to the application of infrared sensitive material, or in other words, to night vision. Zworykin illustrated this with photos of an infrared telescope that had been built at RCA, and also of an infrared microscope. This was a common light microscope with infrared illumination. A photoelectric device would convert the infrared image from the eye piece of the microscope into an electron image on a fluorescent screen. To demonstrate the quality of the infrared imaging system, Zworykin also presented two stills from movies, which had been illuminated with infrared. One still showed Mickey and Minnie Mouse running from the office of Railroad President Dunkwasser. The name Dunk[el]wasser is probably meant to be German for "Dark Water." The second infrared still showed an armed policeman and the cloud of an explosion. One might wonder whether Zworykin had intended to give off a political message with these two remarkable images.

3. UNITED KINGDOM

Also the British history of electron microscopy contains direct links to Berlin right from the beginning. On 5 October 1934, Louis Claude Martin, assistant professor of technical optics at the Imperial College of Science and Technology, South Kensington, London, wrote a letter to Ernst Ruska, asking:

> "Dear Herr Ruska, I have followed with great interest your excellent experimental work in connection with the electron microscope and, with my research students, I have also been making some experiments in these lines. I have been requested to contribute a short article to the periodical "Science Progress", and I should be glad if you could give me permission to reproduce one or two of the diagrams from your papers in the Zeitschrift fur Physik, of course with due acknowledgements."

The letter continues a little further and ends with a post scriptum:

> "I should be very grateful also if you could spare any actual photograph of a natural object such as a cotton fibre. I do not, however, want to reproduce such a picture." [11]

Maybe this final remark is the most significant, as it illustrates that straight from the beginning on Martin belonged to the very first people who wanted to believe in a biological application of electron microscopy. The fact that he does not want to reproduce the picture, probably indicates that he was aware that better results had to be obtained first.

The article that Martin had referred to, appeared in *Science Progress* in 1935 (Martin, 1935). To a certain extent, it paralleled Marton's first article from 1933 in the Dutch Journal of Mathematics and Physics (Marton & Nuyens, 1933). Martin, too, starts with a general discourse on electron optics, describing the electric and the magnetic lens, and even paying some attention to the focussing properties of space charge, followed by a section on "Applications and Possibilities", in which Martin reveals his strong preference for the transmission type electron microscope, just as Marton had done. Already in the first sentence of his section on applications, Martin writes:

> "While considerations of electron optics have had important applications in the design of cathode ray oscillographs, the most attractive possibilities appear in mi-

[11] Personal archive of Ernst Ruska.

croscopy. Very high magnifications can be given by the use of successive 'lenses'. Fig. 9 shows an electron microscope of the type used by Knoll and Ruska." (Martin, 1935)

This Figure 9 was reproduced from Knoll and Ruska's article *Das Elektronenmikroskop*, submitted in June 1932 (Knoll & Ruska, 1932b). About their electron microscope he remarks:

"So far, Ruska and his colleagues have only studied simple objects like metal foils showing small apertures, also carbonised cotton fibres and the like, but it will probably be possible to study biological preparations after suitable "staining" and mounting methods have been developed." (Martin, 1935)

This remark is clearly a reference to the "manual", that Ruska had submitted in December 1933 (Ruska, 1934a), and which is mentioned shortly afterwards by Martin. The phrase "Ruska and his colleagues" is quite significant in itself. At the same time, it is rather surprising that Martin does not make any reference to Ladislaus Marton, who had already published about "suitable staining" in the London journal *Nature* a year before.

In his article Martin continues with the issue of calculating the resolving power of the electron microscope. On the basis of Knoll and Ruska's formula for the electron wavelength,[12] he finds a resolution of 1.6 Å for an acceleration voltage of only 1.5 kV. This outcome is based on a far larger aperture than usual, since Martin believed still a lot could be done about the optical aberrations that oblige to use a very small aperture. At the end of the article Martin does not hesitate to conclude:

"In bacteriology there is ground for hope that even filter-passer organisms, hardly amenable to study by ordinary optical methods, may ultimately be 'brought to book', since the limits of resolution set by the finite wave-length of ordinary light can now be surpassed." (Martin, 1935)

The term "filter-passer organisms" was the common denomination in the 1930s for organisms that could pass through unglazed porcelain, that is, for viruses and bacteriophages. In fact, this makes Martin the first ever to have made a reference to the idea to study viruses with an electron microscope.

Martin's focus on a low acceleration voltage was certainly a deliberate choice. In the introduction of the paper he remarks that he will only discuss electron velocities that are small in comparison to light, apparently wishing to avoid the need to take the effects of the relativity theory into account as well. Also when it comes to the theory of electron lenses, he proved to be well-informed. He explains that magnetic lenses cannot be concave, based

[12] $\lambda = 10^{-8} * \sqrt{(150/V)}$.

on the formulas which had been given by Johannson and Scherzer (1933). For Davisson and Calbick's formula of the focal distance of an electric lens, he points out the fact that the outcome could be a negative value, but adds: "It is not a complete formula as it stands" (Martin, 1935).

Martin's first paper on electron optics is seldom quoted. Nevertheless it is worth to be remembered, since from here on another advocate of biological electron microscopy had stood up and joined the small party which already consisted of Ruska and von Borries, Marton, Anderson and Fitzsimmons, Scott and McMillen and Burton and Hall. Another member at that time, Heinz Müller, we will meet in the next section.

On 8 August 1936, Martin and his co-workers Vaughan Whelpton and Derrick Parnum, submitted to the *Journal of Scientific Instruments* an article called "A new electron microscope", which is generally known as the paper about the first British electron microscope (Martin, Whelpton, & Parnum, 1937). Just like Walter Knecht (Knecht, 1934), Martin had wanted to combine electron microscopy with light microscopy. The arguments for this were understandable:

> *"It appeared to the writers that any practical application of electron image formation to microscopical problems would be dependent on the possibility of observing objects first by an optical microscope and then by the electron system, so that one could step from the known to the unknown by selecting particular object details optically for examination by the lower resolving limit." (Martin et al., 1937)*

The idea was given shape by building an electron microscope and attaching a self-made light microscope to its side. A big circular box connected the two instruments, as shown in Fig. 2. A round disk inside this box served as holder for a maximum of six objects. By rotating the disk, objects could be moved from the light beam to the electron beam. This construction required extremely precise engineering. You just have to realise that the holes in the object-holder for the beams to pass through were only 0.008 inch (0.2 mm) wide, hardly visible to the human eye. It will be clear that the rotating device had to be operating very smoothly to have these holes end up at the right place each time. Fortunately Martin had found help outside the Imperial College. In the acknowledgements it says:

> *"The authors are greatly indebted to Mr A.P.M. Fleming for his encouragement in the development of the present apparatus, the construction of which would have been all but impossible without the assistance of the Metropolitan-Vickers Electrical Co., Ltd., who prepared the final working drawings, and undertook the actual manufacture of the instrument at their own expense." (Martin et al., 1937)*

The later Sir Arthur Fleming was director of research at Metropolitan-Vickers at the time. He had contributed 400 British pounds from his

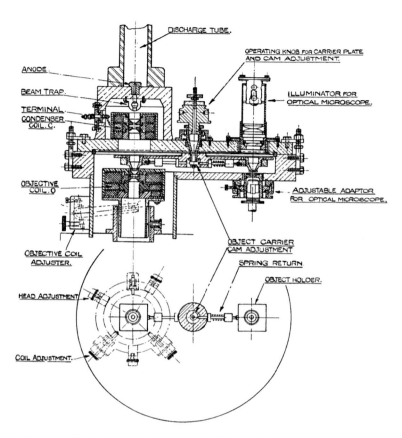

Figure 2 Left half is electron microscope and right half is light microscope. Path of electron beam and path of light are vertical, coming from the top. Image from Louis C. Martin, R. Vaughan Whelpton and Derrick H. Parnum, A new electron microscope, *Journal of Scientific Instruments* 14 (1937) 14–24, Fig. 3.

research budget. Another 800 pounds had been supplied by the Royal Society and the national Department of Science and Industrial Research (Mulvey, 1985). At the time the rate of the pound was five US dollars, or twenty German marks. Because of this involvement of Metropolitan-Vickers, some believe that this microscope was the very first industrial instrument ever, but such a claim seems to be a bit farfetched. It was an experimental design that did not really work out very well in the end (Mulvey, 1985), it was partly made in commission, and never taken into serial production. At the other hand, the company would eventually start producing electron microscopes after World War II, and that commercial development was related to the fact that they already had built a microscope in 1936 (Mulvey, 1985).

The electron optical part of Britain's first electron microscope was based on Ruska's "successful arrangement," as the authors call it, and worked with a 20 kV generator (Martin et al., 1937; they refer to Ruska, 1934a). In several respects, it was improved though, since it was suited for electron diffraction, it made use of internal photography and it provided room for a fluorescent screen as well internal photography of the intermediate image. Additional magnetic coils compensated for the earth's magnetic field, and could also be used to deflect the electron beam. The designers had even been concerned with the aesthetics of these deflecting coils:

> "All the coils are mounted in brass sheet bent into the form of a half-tube, so that their appearance does not detract from that of the rest of the apparatus." (Martin et al., 1937)

The first result obtained with the instrument was the 780-fold magnification of a piece of resistance wire, just 0.0008 inch (0.02 mm or 20 µm) thick. Another image showed that at least details of 1 µm could be distinguished. As a first result, this will have been satisfactory, but it did not surpass the light microscopical resolution.

It is interesting to see that Martin, Whelpton and Parnum's paper gives outspoken support to the thesis of two electron microscope schools. The first sentence says:

> "The term 'electron microscope' is at present used to denote two classes of instrument. In the first, the main object is the study of the thermionic emission of the cathode, of which a direct image is formed by suitable electron-optical systems; in the second, an image is formed of some independent object irradiated by the electron beam much as in an ordinary microscope, except that the place of the light is taken by cathode rays. The present instrument is of the second kind, and has been designed with the express object of adapting the new method for practical microscopy." [emphasis by JvG]

This also supports the view that Martin and his colleagues belong to the transmission school. In contrast to Marton, however, they have no problem at all to acknowledge Ruska's central role:

> "The pioneer steps in the construction of instruments of the present kind were taken by Knoll and Ruska, and were continued by Ruska working alone." (Martin et al., 1937)

This remark must have been pleasant for Ruska, but will have hurt von Borries, if he ever happened to read it, which he will no doubt have done at some stage.

4. GERMANY

4.1 The Determination of Ernst Ruska and Bodo von Borries

When Martin, Whelpton and Parnum wrote these kind words about Ruska somewhere halfway 1936, Ruska himself had drifted away from electron microscopy further away than would ever happen again. As mentioned in van Gorkom (2018), Ernst Ruska had officially joined the Berlin television company *Fernseh AG* on 1 December 1933, while his friend and research partner Bodo von Borries had taken up a position with the electricity company *Rheinisch-Westfälische Elektrizitätswerke* in Essen already on 1 March 1933. Although this had brought an end to their involvement with the actual development of a high voltage magnetic transmission electron microscope, it did not mean that they had distanced themselves from it yet. After Ruska's final paper on the polepiece lens had appeared in the 15 May 1934 issue of *Zeitschrift für Physik*, he was invited to give a lecture to the Electrotechnical Association of Berlin on 29 May 1934. On the same day, von Borries had an interview with Siemens-Schuckert Werke in Berlin, where he was offered a job as laboratory manager for the development and monitoring of surge measuring devices from 1 July 1934 onwards. According to Ruska, this had been a conscious move to start a campaign to promote the further development of their microscope, as Ruska told in 1985:

> "Both of us were determined to try whether we could still develop electron microscopy. And then I have prompted von Borries to give up his position with Rheinisch-Westfälischen Werke and come to Berlin, so we could stay in touch more easily, with regard to our mutual actions. (...) Then from both these positions in Berlin, where we could see each other each day, we could undertake our actions to raise money." (van Gorkom & de Haas, 1985; see also Ruska, 1979)

According to Hedwig von Borries—later wife of Bodo and sister of Ernst—Bodo's diary reveals that from July onwards, he and Ruska worked together on 38 days outside working hours in 1934 and some 70 evenings in 1935—primarily on patents that they were going to file in December 1934 and in April and May 1935 (von Borries, 1991).

July also provided for another less expected, but just as welcome event. By the end of the month Ruska received a letter, dated the 23rd, from Heinz Müller, a student of Professor Matthias at the High Tension Lab.[13]

---

[13] Personal archive of Ernst Ruska.

He requested copies of Ruska's articles, since he had been allowed to do a *Studienarbeit* (research assignment) on the electron microscope. It must have been an odd experience for Ruska. Nearly six years before, he had started his own *Studienarbeit* on testing the validity of the lens theory of Hans Busch. Now his supermicroscope was in the hands of a new student, while he himself had been forced to abandon the development. Müller was working together with another student, Eberhard Driest, but this Driest was not going to play a role of any significance, apart from the fact that he would become the first author of a much cited article with Müller. As Ruska commented in 1985:

> "Driest and Müller were also students of Matthias and when I had left, and my apparatus was available, they made efforts to improve the apparatus. They built in, for example, a provisional photo airlock which I did not have; I only had a viewing screen. I had to make photographs of the viewing screen from the outside (...). Anyway, Müller had contacted me right away and had asked for literature and help (...). It appears that Driest was not very interested. Müller, however, was. From then on Müller stayed in contact with me, and of course then I also learned to know him, but I can't recall that I ever saw Driest or that we had a meeting." (van Gorkom & de Haas, 1985)

No doubt, it will have pleased Ruska and von Borries to know that at least something was done with the old microscope at the Polytechnic's High Tension Lab in Neubabelsberg. It also proves that Professor Matthias felt enough sympathy towards the idea of transmission electron microscopy to allow a follow-up on Ruska's studies. It will also have stimulated von Borries and Ruska to follow through their campaign to seek support for the further development of their transmission microscope, especially since this was not the only positive news. Shortly before, on 4 June, Ruska had received the first letter from Ladislaus Marton in Brussels, and a few months later, on 5 October 1934, Louis Martin sent his letter from London to Ruska.

4.2 The Scepticism of Arnold Sommerfeld and the AEG Research Institute

The description of this phase as an incubation period may seem less applicable, when considering the activities of the AEG Research Institute in 1934. On first sight there is still plenty of activity to be seen with regard to electron microscopy, but a closer look reveals that the development of electrostatic emission microscopy had already entered its final stage. Probably AEG's most noteworthy paper from 1934 is afore-mentioned article by Walter Knecht on "Das kombinierte Licht- und Elektronenmikroskop, seine Eigenschaften und seine Anwendung" (the combined light and electron microscope, its characteristics and application) which had been submitted to *Annalen der Physik* on 26 March 1934 (Knecht, 1934). Knecht described an emission microscope that was fitted with encased magnetic coils at the outside of the tube after Ruska and Knoll (1931). At the inside he had placed an electrical immersion objective after Johannson (1933), and a glass lens. Both immersion and glass lens were attached to hinges, which made it possible to flip either one or the other in front of the cathode and back again. The glass lens was combined with a prism, which would reflect the image in a right angle to a window in the tube wall. With this arrangement, Knecht was able to compare light images with electrostatic and magnetic electron images, made with slow electrons of 0.75 kV. The most important outcome of this study was a large number of photos that showed the similarity between electrical and magnetic images, as well as the advantages of electron images over light images. Of course, these comparisons were very useful to illustrate that reliable and 'true' images can be produced with electron optics.

In general, however, the study of emitting cathodes is a very limited field of research, so it does not surprise us that it had soon been explored to a great extent. Meanwhile, all attempts by the AEG researchers to overcome the chromatic errors of electron lenses failed, as Lin has described in his study already. Lin even gave his Section 4.5 the title: "The development of the electrostatic lenses—a futile search for the achromatic electron lens" (Lin, 1995). In the conclusion of this section he writes:

> "To summarise, one may put it this way: although from the perspective of physics, the Brüche Group started out from a logical point of view, you might say that the complete development of the electrostatic lens and therefore also that of the electrostatic electron microscope ended up on a dead-end street." (Lin, 1995)

At the time, Brüche and colleagues deemed a properly corrected electron lens vital for transmission electron microscopy, since they were convinced that the partial absorption of electrons by the object would cause large differences in electron speed, and therefore serious chromatic errors.

Falk Müller too concludes that AEG did not have much faith anymore in the further development of electron microscopes:

> "Owing to the laboratory's practical focus on the development of electrostatic devices, the construction of the electron microscope was beset with technical problems that could not be solved satisfactory for several years. In addition the electron microscope was merely seen as one of several possible applications of electron optics and was not at the centre of interest. (...) As another reason for the delay in

the construction of the first AEG supermicroscope—particularly put forward in their
later dispute with Siemens—Ramsauer and Brüche mentioned their engagement
on weapon research. In 1934, Ramsauer decided to foster the construction of an
electronic image intensifier rather than to continue the development of electron
microscopes." (Müller, 2009)

This ambivalent attitude towards the electron microscope can also be found in the book *Geometrische Elektronenoptik*, which was published by Ernst Brüche and Otto Scherzer in the second half of 1934. It was an overview of all developments in the field, and was soon hailed by the international physics community for its completeness and thoroughness.[14] With over 330 pages and more than 400 illustrations, the book was another fine example of Brüche's excellent use of corporate communication. Eighty years later it is still a thrill to look through the many pictures in the book.

Near the end, it contains a chapter of no less than 60 pages on *Das Elektronenmikroskop* (Brüche & Scherzer, 1934). Evidently, the emphasis lay on emission microscopy, and the authors also explained why:

"In this chapter we want to deal with electron microscopes and especially the results
of observations. With regard to the results we will highlight the group of cathode
observations, since with its bigger technical importance, it is this field of application
that is already widely accessible with the new method." (Brüche & Scherzer, 1934)

But there are also many references to the papers published by Ruska and Knoll. In general Brüche and Scherzer liked to emphasise that the magnetic microscope at the Technische Hochschule and the electrostatic microscope at the AEG Research Institute had been two simultaneous and independent developments. At the same time, the authors could not avoid making occasional and subtle references to the superiority of the electrostatic instrument, remarking for example:

"All immersion objectives known so far operate without an 'ocular', since very con-
siderable enlargements are achieved with them already; with the system of Fig. 244,
for example, and assuming a microscope length of 1 metre, up to an 800-fold en-
largement can be achieved in just one stage." (Brüche & Scherzer, 1934)
"(. . .) Under the same conditions (length of 1 metre, one stage), the magnetic mi-
croscope can achieve enlargements nearly half as big as the common electrical
immersion objective of 1 millimetre diaphragm diameter." (Brüche & Scherzer, 1934)
"At this stage, the electrical method appears to be easier, since when electron lenses
have to be brought into the vacuum, be it partially or completely, than the sim-
pler built and possibly also gas-freeable electrical lenses cause less complications as

---

[14] See for example: Editorial, Geometrische Elektronenoptik: Grundlagen und Anwendungen, *Nature* 135 (1935) 527.

compared to the magnetic lenses with their windings etcetera." (Brüche & Scherzer, 1934)

Quite interestingly, the authors are well aware that it was Ruska's desire to obtain very high magnifications:

"The wish also to achieve high magnifications in one stage with the magnetic microscope, has led to the application of coils which are built into the tube (. . .). The 1.20 metre tall instrument of Ruska (measured to the cassette at the bottom), which is fitted with such coils, is shown in Fig. 256." (Brüche & Scherzer, 1934)

The image is a photo of the Übermikroskop, which was published in Ruska's 'manual' from 1933 (Ruska, 1934a). The expectations of Brüche and Scherzer are not very high, though:

"The question whether such high magnifications are useful from a physical point of view, will be discussed in Section 6.23." (Brüche & Scherzer, 1934)

This Section 23, called "Current state of the development of the supermicroscope" was the very last one of the chapter and only two pages long. It was preceded by a section of approximately three pages with the rather telling title "The problem of the supermicroscope". First of all, it is explained that in theory, the resolution of an electron microscope could be brought down to atomic distances, depending on the applied acceleration voltage. Subsequently, the authors discuss the theoretical impossibility to observe the motion of molecules, which is a fundamental problem of supermicroscopy according to them. The impact of a single electron, accelerated with 10,000 V, will give an argon atom, for example, an additional speed of 4 kilometres per second:

"It means that the observation of single molecules—apart from other difficulties— will be hopeless." (Brüche & Scherzer, 1934)

Actually, the underlying assumption was that submicroscopical structure would be completely fluid, a concept that will be dealt with in following pages. Moreover, it also implies that the image will be formed by absorption of electrons. In this respect, the authors contradict themselves in the following section, when discussing the results which had been obtained so far:

"The experiments have given two very important results with regard to the high ambitions [emphasis by authors]. First of all, it has been shown that there are objects which can stand (with good heat conduction) the electron radiation necessary for 10,000-fold magnifications, without being destroyed. In the second place, it has been shown that even with these high magnifications, chromatically unadjusted lenses will suffice to image transilluminated foils.

The explanation for both these pleasant conclusions, is the fact that with electron energies of several 10,000 V, absorption and loss of speed fall rapidly against scattering. As a result the electron energy that remains behind in the foil and the consequent heating up, is relatively limited. Moreover, the radiation remains monochromatic to a large extent, implying that the chromatic error of the lens will have no effect." (Brüche & Scherzer, 1934)

We may assume that they had learned this argument from Ruska's 'manual'. Brüche and Scherzer even took the trouble to explain this scattering a bit more. The notion of scattering, however, undid the foregoing fundamental objection to a large extent.

Soon afterwards, another remarkable publication appeared, this time in the 29 November 1934 edition of the *Münchener Medizinische Wochenschrift*. The article was called "About the electron microscope" and was written by Otto Scherzer again, together with his new superior Arnold Sommerfeld (Sommerfeld & Scherzer, 1934). Scherzer had left Brüche and the AEG Research Institute by the end of 1933, to return to the Physics Institute of the Ludwig Maximilian University of Munich, where he had studied with Sommerfeld before (Lin, 1995). As this paper about the electron microscope appeared in a medical journal, the major focus had to be the microscope as a medical tool, in a way envisioned by Ruska and Marton in particular. Therefore, Sommerfeld's involvement is rather surprising, as he was not involved at all with the development of electron microscopes, let alone medical microscopy. He was one of Germany's most famous physicists and had been the teacher of many other well-known physicists, amongst whom were seven Nobel Prize winners (Werner Heisenberg, Wolfgang Pauli, Peter Debye, Hans Bethe, Linus Pauling, Isidor Rabi and Max von Laue) (Forman & Hermanns, 2008). His student Werner Heisenberg, for example, had just received the 1932 Nobel Prize in Physics. However, as Sommerfeld was an expert on quantum physics, he was certainly entitled to speak out on electron physics.

In their article Sommerfeld and Scherzer are quite clear about the origin of the electron microscope. Simultaneously and independently, Brüche and Johannson from AEG's Research Institute had built the electrical one, and Knoll and Ruska from the High Tension Lab of the Technische Hochschule in Berlin had built the magnetic one. Furthermore, they are absolutely positive about the study of cathodes with the emission microscope. But it becomes a different story, when possible biological and medical applications are concerned:

"The question whether biology and medicine may count on enrichment by electron microscopy in the near future, can unfortunately not be answered in such a positive

sense. The observation of live processes is out of the question, as all living matter will be ruptured or dried out in the vacuum pumped research vessel." (Sommerfeld & Scherzer, 1934)

But even if one wished to study dead objects, there would be no hope. About transmission electron microscopy, and alternatively reflection electron microscopy, they say:

"Both these methods have the disadvantage that organic structures will be easily damaged by the impact of the electrons, and that electrons after the passage through the object, and even more after reflecting on the surface, will show considerable differences in speed, which will make it difficult to use them for images. A further problem is that organic matter is usually such a bad conductor that absorbed electrons cannot be diverted fast enough and this will disturb the fields that determine the electron trajectories. On all these grounds a competition between electron and light microscopy is hardly to be expected in the near future, when the imaging of organic structures is concerned." (Sommerfeld & Scherzer, 1934)

In short, differences in electron velocities will cause chromatic aberration, objects will be damaged and they will charge up, three reasons making it impossible to see anything. Assuming that the central question of the article was "whether biology and medicine may count on enrichment", then the answer was a loud and clear negative. But subsequently also Sommerfeld and Scherzer enter into an intriguing self-contradiction, just as Scherzer does together with Brüche:

"Despite all of this, an electron optical image of organic objects is possible, as proven by images 4 and 5." (Sommerfeld & Scherzer, 1934)

Image 4 is Ruska's photograph of a charred cotton fibre that he used to estimate the resolution of his supermicroscope. Image 5 had been provided by Marton and is an image of *Neottia nidus-avis* on aluminium foil, which had been published half a year before in Belgium (Marton, 1934d).[15] Sommerfeld and Scherzer's final conclusion reads:

"One will yet have to temper the wish to get results that surpass those of studying organic matter with light microscopy, until the electron microscope is further developed and until further experience is gathered with the preparation and imaging of organic objects." (Sommerfeld & Scherzer, 1934)

This sentence is carefully phrased, but nevertheless it is somewhat ironical that their major objections are based on the assumption that electrons will be absorbed by the object, while Ruska already mentioned the importance

---

[15] Sommerfeld and Scherzer write that the photo was published in *Nature*, but that is incorrect.

of scattering in the very same article in which he had published the cotton fibre. It is even a bit puzzling that the importance of scattering is discussed extensively by Brüche and Scherzer in *Geometrische Elektronenoptik*, while this argument is omitted in the paper with Sommerfeld.

Altogether, we have two big names now who sided against biological electron microscopy in 1934: the bacteriologist Jules Bordet in Brussels, as we already saw in Section 1, and here Arnold Sommerfeld in Munich. According to Qing Lin, we may even add another well-known name, which is Gustav Hertz. In an interview with Wilhelm Walcher, a former co-worker of Hertz, Lin was told that the electron microscopical experiments of Knoll and Houtermans at the Physics Institute of the Technische Hochschule had eventually been terminated, because of Hertz' scepticism and lack of interest. It could be argued that Knoll and Houtermans' studies concerned emission microscopy, but then this still means that Hertz did not believe in early electron microscopy—not even the successful version of it (Lin, 1995).

4.3 Campaign for a Biological Transmission Electron Microscopy

It is not known whether Ruska and von Borries had been immediately aware of Sommerfeld's comments, but if they were, it did not deter them from filing a total of eight patents from December 1934 onwards. They did so separately—half of the patents were submitted by Ruska and the other half by von Borries. Hedwig von Borries has later explained to Qing Lin that they had done this to avoid their patents being labelled as shop inventions by their employers (Lin, 1995). Ruska patented those ideas that had little to do with electron optics, as he was working for a television company, while von Borries could patent electron optical ideas as far as they did not relate to high-voltage technology. The explanation appears to make sense, also for the reason that the application dates show signs of mutual orchestration. On 6 December 1934 von Borries filed an application for a device to irradiate a surface from the same angle as it is imaged (von Borries, 1934), and on 11 and 13 December Ruska filed the two patents that were already mentioned in Section 1, when discussing Marton's microscope no. 3. One was an airlock to insert photographic plates into the vacuum, the other one was an airlock for objects (Ruska, 1934b, 1934c). The remaining five patent applications would follow some time later, in April and

May 1935.[16] Meanwhile, on 12 December, von Borries also delivered a lecture on electron microscopy to *Haus der Technik* in his hometown Essen, which illustrates once more that von Borries and Ruska were determined to somehow continue the further development of the instrument.

At the same time, the new students Driest and Müller were ready to publish the results of their work with Ruska's old Übermikroskop in Berlin-Neubabelsberg, but immediately received their share of scientific scepticism. The two had managed to make photographs of tiny hairs on the wings and legs of *Musca domestica* (the housefly)—no doubt an object which is not that difficult to lay hands on in a physics laboratory. These were the first electron microscopical images of animal tissue ever, which provided new evidence that Ruska's Übermikroskop was capable of surpassing the resolution limit of the light microscope. This will explain why they sent their images to the most important of German journals, *Die Naturwissenschaften*. Only half a year before, the just as prestigious English journal *Nature* had published Marton's far less convincing photograph of charred sundew cells. On 8 January 1935, chief-editor Arnold Berliner replied with the following three sentences:

> *"Dear Herr Müller, Publishing the electron optical enlargements in* Die Naturwissenschaften *does not really make sense. Maybe* Die Koralle *or* Die Umschau *would be suited. Therefore I can not use the manuscript."*[17]

Die Koralle was a "monthly magazine for all friends of nature and engineering", as it called itself. *Die Umschau* was a similar non-specialist science magazine. If this answer was not intended as an insult, at least it looked like one, while Berliner happened to be not just anybody. He had a PhD-degree in physics, had been collaborator of Emil Rathenau, the founder of AEG, and had been director of AEG's light bulbs factory. On a trip to the US he had learned to know Thomas Alva Edison personally, had been a personal friend of composer Gustav Mahler and was till his death a close friend of physicist Max von Laue, who would later write Berliner's obituary. From that eulogy, it can be learned that AEG sacked Berliner in 1912, since he and Rathenau never got on very well. He then managed to convince the Berlin publishing house Springer to start a journal that had to become

---

[16] Only two of them were granted: Ernst Ruska, Kreisringförmige Polschuhe für magnetische Elektronenlinsen, *German Patent 730719*, priority date of 27 April 1935 and granted on 17 December 1942; Bodo von Borries, Achromatische Linse für die Abbildung mit Elektronenstrahlen, *German Patent 721417*, priority date of 11 May 1935 and granted on 30 April 1942.

[17] Personal archive of Ernst Ruska.

the German counterpart of the British journal *Nature*. The name of the new journal, *Die Naturwissenschaften* is certainly no coincidence therefore. Berliner managed to make it one of Germany's most prestigious journals indeed, thanks as well to his particular interest in optics and biology—a detail that makes his reply to Heinz Müller even more remarkable. It should be mentioned, however, that Berliner's life had become difficult in 1935, as he was from Jewish descent. That same year, Springer was forced by the Nazis to fire him. He was already 72 then and in the following years he would hardly leave his home anymore. When the Nazis wanted to drive him out of his house as well, he took his own life on 22 March 1942 (von Laue, 1946).

On 16 January 1935, a week after Berliner's disenchanting answer, Driest and Müller submitted their article to the less illustrious *Zeitschrift für wissenschaftliche Mikroskopie und mikroskopische Technik*. There it appeared with the title "Electron microscopical images (electron micrograms) of chitin objects" (Driest & Müller, 1935). Chitin is the substance of which the external skeleton of insects is made. As the title already indicates, the article consists mainly of images. One of them shows a magnification of 5000 times of hairs on a leg. A detail of it is further enlarged 5 times by photographic means, resulting in a magnification of 25,000 times and showing a very thin hairlike shape estimated to be 0.04 μm (400 Å) wide. This is 5 times better than a common light microscope will do. In the case of Ruska's cotton fibre, his way of determining the resolution had left room for doubt whether he had surpassed the limit of light microscopy. This little hair of 0.04 μm, however, could hardly be disputed. Another interesting aspect of the article are the first signs of the birth of a scientific community. Driest and Müller could refer to three of Marton's papers from 1934, showing that they were not the only ones in this field. Some time later, Marton would start referring back to them (Marton, 1936a, 1936b).

On the same day that Driest and Müller submitted their article, von Borries gave another talk on electron microscopy, this time to the German Physical Society, at a meeting in Hannover. And in the same month, an English translation appeared of Ruska's canvassing article "The electron microscope as supermicroscope", which he had written just before leaving the Technische Hochschule by the end of 1933 (Ruska, 1934d). The translation was published in *Research and Progress*, which happened to be the English edition of the German journal *Forschungen und Fortschritte*. The translation matches the German original exactly and therefore it also ends with Ruska's first direct appeal to biologists and physicians to join the development of electron microscopy (Ruska, 1935a). By coincidence, this English translation appeared at the same time as Louis Martin's paper in the British journal *Science Progress* (Martin, 1935).

By now, von Borries and Ruska were going to bring into position two additional features of their marketing campaign, as you might call it. Apart from writing articles, giving talks and collecting a number of patents, they also wanted to gain support from established scientists in an attempt to interest industry in the commercial development of electron microscopes. Their first appointment was on 29 January 1935 at the Kaiser-Wilhelm-Institut für Biologie in Berlin-Dahlem. The Kaiser-Wilhelm Institutes belonged to the Kaiser-Wilhelm-Gesellschaft, a highly respected state institution, harbouring numerous famous scholars, although many of them had been forced to leave or motivated to do so by the Nazis. At the Institute for Biology in Dahlem, Ruska and von Borries were received by its director Fritz von Wettstein and his assistant Georg Melchers. Both of them are mentioned in Robert Olby's thorough treatise "The Path to the Double Helix" (Olby, 1994) and from there it becomes clear that Ruska and von Borries had addressed exactly the wrong people.

Von Wettstein and Melchers were so-called physiological geneticists, who were working out the link between genetic traits like the colour of petals and the physiology that determines these traits. It was known for quite some time that such traits were regulated by the presences of enzymes. These enzymes cause biochemical changes, which affect all sorts of traits. The molecular genetics of these enzymes was completely unknown, so it was even considered possible that these enzymes might be the actual genes themselves—instead of the genes' products, as we know now. It meant that Von Wettstein and Melchers focussed their attention mainly on enzymes and their target molecules, and these were believed to move freely within the cell. However, if the concept of a specific spatial distribution of molecules does not exist, and nor is there a desire to introduce one, it will be of no interest to determine at which specific spot a particular molecule is present at some moment in time; such information will be completely superfluous. Also the suggestion to study intracellular structures would not work. In 1935 the conviction was still widespread that the maximum size of molecules was quite limited. The existence of macromolecules had only been suggested some ten years before by Hermann Staudinger, and was still a young and disputed concept (Olby, 1994).[18] Many stuck to the idea that

---

[18] Staudinger received in 1953 the Nobel Prize in Chemistry "for his discoveries in the field of macromolecular chemistry" (www.nobelprize.org).

bigger structures were so-called colloids: these were rather loose arrangements of smaller molecules, being held together by weak atomic forces. This view was even embodied in a discipline called colloid chemistry. Colloid chemists assumed that big structures would be prone to rapid change as well. So any attempt to find meaningful structures inside a cell, was just as useless as trying to make enzymes visible.

Despite all this, Wettstein and Melchers were willing to spend two hours on discussing the issue (van Gorkom & de Haas, 1985). Nor had they any reason to be explicitly negative towards Ruska and von Borries. The latters' request to the Kaiser Wilhelm Institute for Biology was only to give support to the view that electron microscopy would open up new fields of research. Unfortunately Ruska and von Borries were no serious party in a discussion about physiological chemistry. As Ruska recalled the unfavourable arguments of their discussion partners in 1985:

> "I did not understand a thing of their ideas. But I got the impression that these people thought that there are no fixed structures below [the limit of] what you can see with a light microscope; there everything is moving, everything is fluid. There are no structures anymore from which you can learn anything sensible. And also this opinion is certainly forgivable, because biological objects contain water, and you have this Brownian motion of molecules. One could certainly believe that there are no detectable, fixed, fine structures." (van Gorkom & de Haas, 1985)

If von Borries and Ruska had been trained in physiology, they would have known that not all experts believed that permanent submicroscopical structures could not exist. Maybe the most explicit example of a different perception comes from neuroanatomy. Already in 1898, the Italian Camillo Golgi had described the "endocellular reticular apparatus", widely known nowadays as the Golgi apparatus, which is a complex structure of membranes that he had discovered in the interior of nerve cells (Golgi, 1898; see also Golgi, 1967).[19] And as early as 1900, Albrecht Bethe made drawings of neurofibrils, which are intracellular structures of nerve cells as well (Bethe, 1903). Ruska and von Borries were no biologists or chemists, though, and that was exactly the reason why they sought the support of scientists like von Wettstein and Melchers. Fortunately for them, there were also people like Ernst's brother Helmut, who did know about intracellular structures,

---

[19] The Golgi apparatus can be found in practically all eukaryotic cells. Golgi received in 1906 the Nobel Prize in Physiology or Medicine together with Santiago Ramón y Cajal "in recognition of their work on the structure of the nervous system" (www.nobelprize.org).

and therefore they did not give up immediately after this first fruitless attempt.

A complete list of actions by Ruska and von Borries can be found as Appendix E in the back of Ruska's historical review "The early development of electron lenses and electron microscopy", which he published in 1979 (Ruska, 1979). It shows that three days after the meeting with von Wettstein and Melchers, von Borries wrote a letter to Dr. Gajewski, a board member of the IG Farben establishment in Wolfen. In it he asked whether IG Farben might be interested to apply electron microscopy to fibre research. After an initial request for more details, they finally answered that they did not see how electron microscopy could benefit them. They advised to contact the IG Farben establishment in Oppau, but it turned out that Oppau was also not interested.

On 20 February 1935 they approached one of Germany's most celebrated physicists, Max Planck, who is probably best known for Planck's constant h, which can be found, for example, in de Broglie's wavelength equation $\lambda = h/mv$.[20] In 1935 Max Planck was already 77 years old and president of the *Kaiser-Wilhelm-Gesellschaft*—the institute that would even be renamed in Max-Planck-Gesellschaft after World War II. Planck was willing to receive them three months later, on 21 May. On 23 February Ruska wrote to Dr. Harting, board member of the Carl Zeiss company in Jena, the famous microscope factory of which Ernst Abbe had once been co-owner, together with Carl Zeiss himself. This letter also resulted in an appointment on 21 May.

On 12 March Ruska approached Askania, the company where his elder brother Walter worked, and on 21 March Fernseh AG, his own employer, and Zeiss Ikon, the camera factory founded by Carl Zeiss. These last three attempts gave no results. Von Borries, finally, managed to get in contact with Krupp, the large steel manufacturer. On 12 March he had written to Dr. Goerens, a board member of the Essen branch, whether electron microscopes could be of any interest to the steel industry. It is not unlikely that he already knew Dr. Goerens somehow, as both of them had ties to Essen, while von Borries had family ties to the steel industry as well. This contact resulted in some further talks with Krupp, and a final offer by Dr. Goerens to assist at negotiations with the Kaiser-Wilhelm-Gesellschaft. Apparently this help never materialised. So, to summarise, in the end it all came down to the two appointments on 21 May: one with Planck and the other one

---

[20] Value of h is $6.62606896 \times 10^{-34}$ Js.

with Zeiss Jena. When we asked Ruska how they had selected companies, the answer was very simple:

"We looked for people with money. And with an interest in science. We had certainly not started by selecting the most suitable places, but picked places where we happened to know someone or to which we had ties." (van Gorkom & de Haas, 1985)

Meanwhile, drawing public attention continued as well. On 18 March 1935 von Borries gave in Düsseldorf a third talk on electron microscopy, and a month later, on 27 April, a new article appeared which was written by him and Ruska together. It was their first joint paper since the article on the transillumination of metal foils, submitted in April 1933. This lapse of two years seems to illustrate the impression that the two of them had gone through a period of chill after von Borries' move to Essen and the subsequent solitary actions by Ruska in late 1933, as described in van Gorkom (2018). If true, problems had clearly been sorted out now and their renewed spirit had resulted in a contribution to *Zeitschrift des Vereines deutscher Ingenieure* (also known as *VDI-Zeitschrift*) (von Borries & Ruska, 1935). The paper was titled "The electron microscope and its applications", which was identical with the title of Marton's simultaneous article in the French journal *Revue d'Optique* (Marton, 1935c).

Like Marton they explain the differences between the electrical and magnetic electron microscope, and although they do not say that the two different systems had been developed "simultaneously and independently", as Brüche, Scherzer and Sommerfeld presented it, they have their own way of demarcating territory. They do so by defining four types of electron microscopy: self-emission, emission by foreign substances, reflection and transmission. The first type referred to the classic imaging of cathodes and the second type to the emission by substances that are mounted on the cathode. For both types, they made references to Brüche, although they took the trouble to stress that mounting foreign substances had originally been the idea of Fritz Houtermans, just as Knoll and Ruska had already pointed out three years before (Knoll & Ruska, 1932b). The electrical microscope was well suited for these two sorts of electron microscopy, especially since you will not need very high acceleration voltages. For reflection and transmission microscopy, however, you do need high acceleration voltages, they say, and then you have to use a magnetic microscope instead, as electrical lenses do not work very well under such conditions. These were prophetic words, which were going to be heavily disputed by AEG in following years.

Another immediate parallel with Marton's paper was the presentation of an airlock for objects, as well as the suggestion to deflect the electron beam in order to reduce the time that the object is exposed to it. As mentioned before in van Gorkom (2018), the latter is one of the rare cases that Marton has been clearly the first to suggest a technical improvement. Marton is referred to by von Borries and Ruska, when discussing the biological application of the instrument, although they only mention Marton's first paper, which he had published in *Nature* nearly a year before. It is not clear why they do not refer to any other of Marton's papers from 1934. They also showed two of the images of housefly hairs, made by Driest and Müller, and state that the resolution of 0.04 μm, which had been obtained by Driest and Müller, was the best achievement of the electron microscope so far. It is another interesting example of the new possibility to do mutual name dropping. Subsequently, the authors felt tempted to speculate on imaging living objects by using Lenard windows. As has been discussed already, they were fair enough to acknowledge that Reinhold Rüdenberg had been the first to mention this in one of the Siemens patents.

In spring 1935 interesting news came from London as well. The 6 April issue of *Nature* had printed a review of Brüche and Scherzer's new book *Geometrische Elektronenoptik*.[21] The reviewer's use of quotation marks to emphasise new terminology shows that geometric electron optics was still considered a novelty:

> *"Electrons behave in many respects like light. They may travel in straight lines; may be 'refracted' in electric or magnetic fields, may be focussed as by a lens in suitably graded fields, or may be caused to produce interference patterns in properly disposed apparatus."*

The terms "electron lenses" and "electron microscope" were emphasised as well. The review is quite favourable, also about electron microscopy:

> *"A very interesting set of comparison photographs of the same objects through the ordinary microscope and the new electron microscope allow the reader to form his own conclusions about the present state of the new technique in relation to the old."*

The comparison photographs in the book actually show cathode surfaces and therefore relate to emission microscopy, but this cannot be deduced from the review.

Finally the day came that von Borries and Ruska were going to have their two important meetings. The first meeting on 21 May 1935 was with Max Planck. As can be learned from the list of actions in Ruska's book,

---

[21] Editorial, Geometrische Elektronenoptik: Grundlagen und Anwendungen, *Nature* 135 (1935) 527.

he and von Borries tried to convince Planck that the involvement of his Kaiser-Wilhelm-Gesellschaft was desirable for the further development of electron microscopy, and outlined the possible far-reaching consequences of results known so far. The other meeting was with Dr. Harting and Prof. Dr. Bauersfeld of the Carl Zeiss firm in which they discussed the development of electron microscopes and cathode-ray oscilloscopes. The two gentlemen turned out to be interested, which will have caused considerable excitement with the two young inventors. In the next days they started to make detailed plans with Zeiss for setting up a laboratory for the development of electron microscopes and oscilloscopes as part of an existing Kaiser-Wilhelm-Institute. On 29 May Ruska even sent proposals for contracts, in case Zeiss was going to hire them to run the new facility. On 3 June they discussed with Zeiss issues concerning patents. On 4 June Max Planck provided a written recommendation to help them win a director of one of the Kaiser-Wilhelm-Institutes for their plans. On 6 June they sent additional documentation concerning relevant patents to Zeiss. This also included the Siemens patent filed by Rüdenberg. And so it happened that on 13 June 1935 Zeiss cancelled the negotiations, because of this Siemens patent (Ruska, 1979).

The disappointment must have been immense. It will hardly be a surprise that Ruska's activity list shows no more entries from July 1935 till September 1936. When we asked him in 1985 about this time gap of fourteen months, it turned out that he had even repressed the memory of it, as we were the first to draw his attention to it. Until then he had often stressed that he and von Borries had strived for three years to realise the further development of the electron microscope, during this period from 1934 till 1937 (see for example Ruska, 1970, 1974; Lambert & Mulvey, 1996). Then, Ruska's wife Irmela, who was also present during the interview, gave a helping hand by exclaiming:

> *"But Ernst, this is exactly the time that I was in Berlin!",*

which was a reference to their first engagement. Ruska acknowledged:

> *"Yes, perhaps I learned to know my wife in this period. Could be. (. . .)*
> *It may be that after we had made such an effort, that after the cancellation by Zeiss,*
> *we had become a bit numb. It can very well be that we did not undertake much*
> *anymore for some time. It is possible." (van Gorkom & de Haas, 1985)*

In later years, Hedwig von Borries and Qing Lin have also drawn attention to this time gap (von Borries, 1991; Lin, 1995). So, illusions about a brilliant future had just been scattered, but finally love was in the air. The failed talks with Zeiss were not the final curtain for Ruska's old electron microscope, however. After Müller and Driest had finished their assignment at the Technische Hochschule, the instrument was given to a new student, Friedrich Krause. He started after the summer break of 1935 and was going to make some important contributions in the next year.

4.4 Reinhold Rüdenberg

The dramatic failure to set up a laboratory with Zeiss and the Kaiser-Wilhelm-Gesellschaft might explain why Ruska decided to finally write a letter to Reinhold Rüdenberg. It was he after all who had been responsible for the Siemens patent which had caused the breakdown. On 22 July 1935, Ruska wrote:

"Dear Herr Professor, The Austrian patent 137 611 and the French patent 737 716, both regarding the electron microscope, became known to me by chance. (. . .)
I assume that you are informed about the fact that I built electron microscopes and carried out investigations with them in the High Voltage Laboratory of the Technological University, Berlin, before the priority date of your first German application. Since that time I have continued to work in this field, and I am still interested in the development of this project in the future. Naturally, I do not want my hands possibly to be tied at some later date because of so far-reaching protection rights as are represented by the two patent documents.
I believe it to be appropriate that prior to the possible occurrence of concrete cases, there should be an understanding with me about the right to use jointly a patent, to which I am entitled. (. . .)" (Ruska, 1986)

Rüdenberg, who still worked at Siemens then, answered on 29 July:

"In reply to your kind letter of 22.7.1935 I have asked our Patent Department to compile the respective material first. I myself will leave for a holiday in the next days and I will be away for several weeks, so it will not be until the beginning of September that I can personally deal with the matter. Should it be your wish to do anything in the mean time, I kindly ask you to turn to Dipl. Ing. Wolf of our Patent Department." (Ruska, 1986)

Ruska wrote to Wolf immediately, but received no reply. Quite surprisingly, Gunther and Paul Rudenberg found in their father's and grandfather's diary an entry for Monday morning 16 September 1935, saying "10 Dr. Ruska Elektr.Mikrosk." Ruska has never made any mention of this appointment and neither do Gunther and Paul Rudenberg know whether the meeting ever took place.

Rüdenberg would eventually flee Nazi Germany half a year later, on 2 April 1936, a train taking him to Amsterdam, where his wife Lily and youngest son Hermann joined him the next day in the Victoria Hotel, opposite Amsterdam Central Station. From there they travelled on to Lon–

don where they were reunited with the two other children, Angelika and Gunther, who had been attending boarding schools in France and England (Rudenberg & Rudenberg, 2010). According to the Rudenberg family, "The Siemens firm graciously, and perhaps with some risk, packed up the family belongings at their Charlottenburg home and shipped them to England" (Rudenberg & Rudenberg, 2010). On 16 October 1938, all of them arrived safely in New York on the *SS Volendam*.

4.5 Max Knoll

After von Borries and Ruska had finally given up on the transmission electron microscope, as it seems, there was suddenly news from Ruska's old supervisor Max Knoll. He had written a general introduction to the electron microscope for a medical journal called *Zeitschrift für ärztliche Fortbildung* and as such it reminds us of the paper by Sommerfeld and Scherzer in Munich Medical Weekly the year before (Sommerfeld & Scherzer, 1934). Knoll's paper was called *Das Elektronenmikroskop* (Knoll, 1935a), just like his article three years before with Ruska, and appeared in two parts on 15 November and 1 December 1935. Since Knoll had been working in a quite different research field at Telefunken for some time and there is no mention of the company at all, he had certainly written this contribution in a personal capacity. In several respects, his paper looks like an answer to the one by Sommerfeld and Scherzer. On the first page, Knoll already makes clear that he and Ruska had been the first to build an electron microscope. His description of the instrument's construction is introduced in just as outspoken a manner:

> "Since so far solely the magnetic electron microscope has proven to be useful for making large magnifications, only the basic build-up of this microscope will be described here." (Knoll, 1935a)

This did not prevent him, however, from explaining the difference between electrical and magnetic lenses, just as Sommerfeld and Scherzer had done. Next came a discussion of emission microscopy and results so far obtained. Since he had limited himself to magnetic instruments meanwhile, he only referred to his own emission work, done in 1932 together with Fritz Houtermans and Werner Schulze, without naming Brüche or Johannson in this context at all. And just like Sommerfeld and Scherzer again, he concluded his paper with possible applications of transmission microscopy, as was to be expected in a medical journal. There Knoll finally presented three transmission images: Ruska's cotton fibre from his 'manual', the *Neottia* image by Marton that had also been published in *Physical Review* and the

Bulletin, and Driest and Müller's housefly hairs (Ruska, 1934a; Marton, 1934b, 1935a; Driest & Müller, 1935). The world of biological transmission electron microscopists already started to look like a crowd.

Noticeable is Knoll's very neutral approach where the prospects of biological electron microscopy are concerned. He concluded the article with:

"Considering the ease with which organic objects change upon irradiation with electrons in vacuum and their relatively complex structure, it can hardly be expected that the electron microscope will be applied to general medical matters within the next years: its focus will lie much more on the electrophysical and metallographic field. However, the results that have been achieved so far in only three years time, with which the resolving power of the light microscope has been reached for organic study objects as well, are nevertheless so encouraging, that the use of the electron microscope for specific issues in medical research appears to be rewarding by now." (Knoll, 1935a)

Although his conclusion is phrased carefully, it is clear that Knoll does not wish to be as pessimistic as Sommerfeld and Scherzer had been.

Simultaneously with this paper in a medical journal, Knoll had another one published in the *Zeitschrift für technische Physik*, which was going to have far greater significance. It was the text of a lecture given at the eleventh German Conference of Physicists in Stuttgart, which took place from 22 to 28 September 1935. It was called "Charge and secondary emission of electron irradiated objects" (Knoll, 1935b) and it is the first paper ever on scanning electron microscopy. A scanning electron microscope produces an image with electrons that reflect on the surface of the object. The scanning principle is very similar to imaging in television. The image is made with a narrow electron beam which scans the surface of the object line by line, one after the other. The changing intensity of the beam's reflection is registered and this signal is sent to a television screen that copies the reflected lines. In this way three-dimensional looking images can be obtained.

Since Knoll published this study in his capacity of Telefunken employee, the paper was first of all a study of the effect of secondary electron emission. From the company's point of view, secondary emission was a denominator for an annoying problem. Since electrons in a cathode-ray tube would end up at many places where they are not needed, many surfaces inside the tube would charge up, start to emit and disturb the cathode ray. One of the images in Knoll's article shows this secondary emission at the inside of a cathode-ray tube. But at the same time he took the opportunity to study the general behaviour of his new imaging system, based on the scanning principle as described above and using a simple electrical lens. Certainly his most wonderful experiment is a test with a copper electrotype plate, which

Lichtbild des Klischees   *Elektronenbild*

Klischeeabzug

Figure 3 Secondary emission of a halftone image. "Klischee" is printing plate, "Abzug" is print, "Licht" is light," "Bild" is image. From Max Knoll, Aufladepotential und Sekundäremission elektronenbestrahlter Körper, *Zeitschrift für technische Physik* 16 (1935) 467–475, p. 472.

he must have obtained from a printing establishment. In it was etched the portrait of a woman. In this etched surface Knoll had rubbed coal dust, which will show hardly any secondary emission in contrast to the copper details that were protruding from the coal. In this way the image on the plate became already visible to the human eye, but could also be reproduced by scanning it with electrons, as shown in Fig. 3.

Another thing he did, was to show that such an image is enlarged very easily. For this purpose he used the surface of a cathode and commented:

"Without doubt one can see single crystals, while just as in the electron microscope the electron image differs in many respects from the light image."

In an accompanying footnote he discussed the resolving power of the "scanning method" as he called it, and suggested that the scanning method would be more suited for studying crystals, than common electron microscopy. It can therefore hardly be disputed that Knoll knew that he was presenting a new type of electron microscope. The only thing he did not do, was casting its future name.

Unfortunately, also in the case of Knoll's invention of the scanning microscope there is a priority issue. Already in 1927 the German Hugo Stintzing had applied for a patent that describes a scanning device using an electron beam or something comparable (Stintzing, 1927). This patent is titled "Method and device to detect, measure and count particles of any nature, form or size". Stintzing had never attempted to build it. The issue with this patent, however, is that it does describe the scanning principle, but the scan does not produce an image. Instead the output is visualised as a linear graph, so you might say that is only half a scanning microscope, or alternatively, a scanning micrometre, since it does not "watch" (scope, σκοπεῖν), but measures the size and number of particles. Neither can Stintzing be credited for inventing the scanning principle as such. This had already been done in 1884 by the German Paul Nipkow, who came up with the idea to scan an image dot by dot and line by line, and then transmit these lines of dots to a similar machine that puts the dots together again (Nipkow, 1884). Finally, the idea to use a cathode-ray tube to display the transmitted image, came from the Russian Boris Rosing, who patented in 1907 a "device for the electrical remote transmission of images" (Rosing, 1907). Just for the record, it should finally be added here that Rosing happened to be Vladimir Zworykin's teacher, when the latter still lived in St Petersburg.

The medical journal in which Knoll's other article had appeared simultaneously, gives a vivid example of the political situation in Germany by the end of 1935. Halfway between the two parts of his article, a government publication can be found, written by Dr. Gütt, Director of the Ministry of the Interior. He explains to physicians the introduction of a "certificate of healthy marriage", laid down in the *Law for the Protection of the Hereditary Health of the German People* of 18 October 1935 as part of the so-called Nuremberg Laws. The first sentence of the publication reads:

"Going by the notion that strong peoples can have eternal life if they have the will for it and if they safeguard their identity and racial sanity, national socialism has recognized the care for public health as one of the most important tasks of the state, and shown it new ways by directing it in terms of hereditary and racial care." (Gütt, 1935)

The law was intended to complement other new laws like the State Citizens Act and the Act on the Protection of the German Blood and German Honour—also known as Blood Protection Act. Dr. Arthur Julius Gütt would later become a senior SS general.

4.6 A Complete Standstill

In the mean time, it started to become uncomfortably quiet in the newborn field of magnetic transmission electron microscopy. If the years since 1934 had looked like the silence before the storm, there was now—by the end of 1935—good reason to believe that the fresh rains of a fundamentally new research method were drifting off and were never to shower down on biology and medicine. Since von Borries and Ruska's aborted talks with Zeiss, nothing substantial had happened, apart from Knoll's general introduction for physicians. By sheer coincidence, Marton was neither publishing anything on electron microscopy for a full year in the same period. In St. Louis, Missouri, USA, McMillen and Scott were working on a magnetic instrument with one lens at the time, but nobody knew, and strictly taken it was an emission instrument. Cecil Hall was experimenting in Toronto with an emission microscope too, while nobody knew, and so were Fitzsimmons and Anderson at Washington State University working on a full-fledged magnetic transmission microscope that the outside world was never going to see. In London, finally, Martin, Whelpton and Parnum were working on their hybrid light-electron microscope, but they were only to publish about it in August 1936.

Also, Friedrich Krause, the new student who had recently started to experiment with Ruska's instrument in the High Tension Lab, did not appear to be very successful. Right from the beginning, Krause was determined to try to apply the electron microscope to biology and medicine. In 1985 Ruska told us about him:

> *"Krause was just as convinced as we were that electron microscopy would become something important (. . .). Krause was a nephew of the chief assistant of Professor Matthias. His name was Czemper, also Jewish, later he went to Palestine. His nephew Krause was a physician, and had heard from him about the electron microscope and had become interested. A very fanatical man." (van Gorkom & de Haas, 1985)*

At the time, Krause was actually not yet a physician, as he was still studying. The mentioning of the Jewish background of Czemper seems to illustrate that the High Tension Lab was still not Nazi-infected in 1935.

Thanks to the help of his uncle, Krause was able to make some improvements to Ruska's instrument. However, as he was a medical student, it is logical that it was his main interest to make biological images. On 13 January 1936 Ruska received a letter from Czemper with the typescript of Krause's first paper and the request to have a look at it.[22] It appears to have been a histological study, comparable to Marton's, in which he showed his first large magnifications of cells, which he had made in the last quarter of 1935 (this can be learned from Krause, 1937a). It turned out that no journal was willing to publish it, although the paper had been looked at by Adolf Matthias and three other professors. Amongst them was August Köhler, who was professor of microscopy and a staff member of Carl Zeiss AG, and who had also advised Krause on light optical matters.

A few months afterwards, another review appeared of the book *Geometrische Elektronenoptik*, which Ernst Brüche and Otto Scherzer had published in 1934 (Liesegang, 1936). It was published in the same "Journal of scientific microscopy and microscopical technique" that had published Driest and Müller's images of hairs on flies. The reviewer was the respected and versatile German scientist Raphael Liesegang (Beneke, 2004/2006), who had specialised in colloidal chemistry—the discipline that did not believe in very large molecules as we saw before. It is no surprise therefore that Liesegang dedicated half of his short review to the main argument in the book against high resolution transmission electron microscopy:

"(...) Their inventors had assumed at first that one would get much larger magnifications than with light. Much of this is taken back here. The hope to get to the level of molecules fails, as the molecules will be changed too much by the process of imaging itself."

On 4 June 1936, Otto Scherzer was to submit his by far most famous paper, called "About some errors of electron lenses," to the *Zeitschrift für Physik* (Scherzer, 1936). By then he had become professor and director of the theoretical physics department of the Darmstadt Polytechnic—just 27 years old (Lin, 1995). His treatise is divided in several sections, the first of which carries the loud and clear title "The impossibility of the achromatic lens". The very last subsection is similarly named "Unavoidableness of spherical aberration." After this publication it was considered absolutely proven that it is impossible to correct the chromatic and spherical aberrations of by then generally used electrical and magnetic electron lenses. As a consequence, the traditional transmission electron microscope has to be operated with very narrow lens apertures, which reduce the maximum resolution considerably, in order to avoid serious spherical aberrations. This insight would become widely known as Scherzer's Theorem. Of course, this limitation could have been used as an additional argument against

---

[22] Personal archive of Ernst Ruska.

high-resolution transmission microscopy, although Scherzer does not do so himself.

In early June as well, there was finally more promising news for Ruska. His old rival Ernst Brüche had invited him to give a talk at the 12th German Conference of Physicists and Mathematicians in Bad Salzbrunn that was going to be held from 13 till 19 September 1936.[23] Brüche was organising the electron optical part of the conference together with Hans Busch, the grand old man of the electron lens theory. At that time, Ruska and von Borries were just to publish a new article in the same *Zeitschrift des Vereines deutscher Ingenieure* in which they had published a year before. Actually, their contribution was split into two subsequent articles, both called "Applied Electron Optics", which appeared on 15 and 29 August 1936 (von Borries & Ruska, 1936). The first article consisted of a general introduction to the topic, followed by the discussion of four applications: the electron spectroscope, electron diffraction tube, electron microscope and the oscilloscope. The section about the electron microscope is especially noteworthy because of its brevity. It is just one short paragraph on a total of six large pages. The reason for this is given in the last sentence:

"The writers have recently presented a survey on the electron microscope in this journal. As no really new results have become known in the mean time, we refer to this earlier paper." (von Borries & Ruska, 1936)

This earlier paper was the article that they had published more than a year before (von Borries & Ruska, 1935). It illustrates that really nothing was going on in the field at that moment, at least as far as published results were concerned. The second article was longer with nine pages and was largely dedicated to television—doubtless the most exciting application of electron optics at that moment. Not long before, on 22 March 1935, regular television broadcasts had started in Berlin, which could be watched in so-called television pubs. The station would soon become known as *Fernsehsender Paul Nipkow* (television broadcasting station Paul Nipkow), which continued its services until nearly the end of World War II (Hickethier, 2008). Regarding their strong emphasis on a field of engineering that looked most attractive at that moment, one is tempted to believe that von Borries and Ruska had decided not to put all their bets on the electron microscope anymore.

It is absolutely interesting to read this treatise on television, which reads like another manual, this time telling you how a 1936 television receiver

---

[23] Personal archive of Ernst Ruska.

and camera were built. Especially interesting are two sections that are dedicated to the work of two men who are well-known from the history of American television: Philo Taylor Farnsworth and more in particular Vladimir Zworykin. Both of them had invented a television camera, based on different principles, which are discussed extensively by Ruska and von Borries. Subsequently, they also pay attention to the night-vision gear that Zworykin had described half a year before in his talk at the Technische Hochschule. Von Borries and Ruska refer to the publication of this talk, and it is rather likely that at least Ruska will have been in Zworykin's audience, as he was most and for all a television engineer at that moment.

Despite the poor outlook of their main dream, life did not look that bad, though, for the two pioneers in August 1936. Ruska had just been invited for a lecture at the Physicists and Mathematicians Conference, he had started to date his future wife Irmela Geigis, von Borries was dating Ruska's sister Hedwig, and moreover the 12th Olympic Games were in town, which had started on 1 August 1936. The official website of the International Olympic Committee tells at the moment of writing this story:

> "The Berlin Games are best remembered for Adolf Hitler's failed attempt to use them to prove his theories of Aryan racial superiority. As it turned out, the most popular hero of the Games was the African–American sprinter and long jumper Jesse Owens, who won four gold medals in the 100 m, 200 m, 4 × 100 m relay and long jump."
>
> "The 1936 Games were the first to be broadcast on television. Twenty-five television viewing rooms were set up in the Greater Berlin area, allowing the locals to follow the Games free of charge."[24]

Quite remarkably, Ernst Ruska was very actively involved with the Games, since he was overseeing the production of television tubes at Fernseh AG. In 1985 he told:

> "During the Olympiad, I had to make new tubes every night. For the recording of the games a Braun tube that was in working order was always needed. So each night I was at work to oversee the pumping out of the tubes in the oven. These tubes were heated to release the last traces of gas from the inside walls, so they could be pumped out. Therefore they had to be heated to 400–450 degrees [Celsius] for many hours, while they were attached to the pump, and under low pressure. The temperature is not allowed to rise too far, as otherwise they will collapse due to outside air pressure. So, to oversee this pumping out I was at the Fernseh AG every night, and during the afternoon or evening I was in the Olympic Stadium with [girl friend Irmela] to watch the matches." (van Gorkom & de Haas, 1985)

[24] http://www.olympic.org (accessed 4 August 2013).

Later he explained that he oversaw the production of tubes for receiving purposes as well as for sending. In the stadium the receivers were needed to monitor the broadcasts (van Gorkom & de Haas, 1985). The recording of the games was also done by Telefunken, the company where Max Knoll was working. Fernseh AG was present at the stadium with a Farnsworth type of camera. Telefunken recorded with Zworykin's iconoscope.[25]

4.7 A Final Effort

Brüche's invitation to give a talk at the upcoming conference proved to be a new source of inspiration to both Ruska and von Borries. Two weeks after the games, Bodo von Borries undertook action for the first time in fourteen months to find a business partner for the development of electron microscopes. On 2 September he approached his highest boss, Dr. Carl Köttgen, Chairman of the Managing Board of Siemens-Schuckert Werke.

Meanwhile, Ruska was due to give his lecture in Bad Salzbrunn on 15 September. The talk was called "Elektronenmikroskop und Übermikroskop" and was going to contain no new results from the speaker himself for obvious reasons. In three years time, though, Ruska's tone of voice had changed considerably, starting to reveal his rather ironical nature, which was very present 49 years later. In his talk he remarked about emission microscopy:

"It should be questioned whether the numerous, already available studies in this field, which have their value for emission research (. . .) should be regarded as real microscopy. In light optics at least, you would not associate the word microscopy with the study of light sources, but much more with the enlarged imaging of any kind of object by means of light rays." (Ruska, 1937)

A little later he added:

"The true electron microscopy practise which is the most important one now and probably also in future—that is for the study of any kind of object by means of electron rays—is the transmission practice though, just like in light microscopy." (Ruska, 1937)

After this he moved on to the arguments in favour of transmission electron microscopy. The most important one was evidently the better resolving power. In connection with this he came with a rhetoric argument, which might have been prompted by his brother Helmut, who had finished his

---

[25] Deutsches Rundfunkarchiv, *Die Olympischen Spiele 1936 im NS-Rundfunk*, http://www.dra.de (accessed 4 August 2013).

medical training by now and who was going to join Ernst's and Bodo's efforts very soon:

> *"The limit to the resolution of light, however, sets bounds to knowledge. An example is histology, which showed a brilliant development since the 70s of the former [19th] century thanks to the microscope, but has now come to a certain halt, which is partly due to the impossibility of higher resolutions." (Ruska, 1937)*

He illustrated the new possibilities of transmission electron microscopy with the latest images that Marton had enclosed in his recent letter to Ruska of 29 August 1936. It was the electron image of *Neottia* cells on a cracked zapon foil, together with a light image of a similar preparation. Ruska also showed four different light images of rat tissue, dyed in four different ways and therefore displaying four very different structures. It was another rhetoric argument, in this case to invert the objection used for example by bacteriologist Jules Bordet, who had protested that he had "already enough trouble to interpret the images obtained with the light microscope". After all, if you do manage to interpret these very different light images of rat tissue, why is it an objection then to make the same effort for electron images?

Marton's recent images of *Neottia* cells were not the only ones that Ruska could show to his audience. There were also unpublished results that had been obtained by Friedrich Krause with Ruska's old Übermikroskop in Neubabelsberg. From a letter by Ruska to Krause, dated 4 November 1936, it can be deducted that Krause had not been able to give a talk at the conference himself, and had given his images to Ruska instead.[26] Some of them were images of *Schlammteufel* skin cells (*Cryptobranchus alleganiensis*). In English *Schlammteufel* is called Hellbender—a giant salamander from North America. From this salamander the epithelium was taken, that is the top layer of the skin, in this case a thin pellicle only one cell layer thick. Krause had simply put this pellicle on top of a metal grid without any kind of preparation. With this primitive approach he could already produce an image with 900-fold magnification in which cells and smaller details could be clearly recognised (Krause, 1937b). This image was certainly better than Marton's recent *Neottia* image; Krause's micrograph showed more contrast, probably because he had used the fixative potassium dichromate and he had cooled the cells to minus 17 °C. In this way, he could expose them to the electron beam for at least five minutes before they would start to tear. For comparison, Krause also provided light images of the same cell type, which

---

[26] Personal archive of Ernst Ruska.

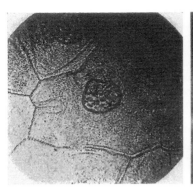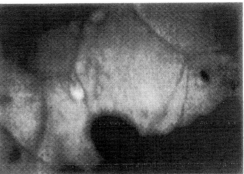

Figure 4 Hellbender cells. Left: light micrograph, magnification 300. Right: electron micrograph, magnification 600. Note the similar details of the membranes in both images. From Friedrich Krause, Neuere Untersuchungen mit dem magnetischen Elektronenmikroskop, in: *Beiträge zur Elektronenoptik*, Leipzig: Johann Ambrosius Barth, 1937, 55–61, pp. 59 and 60.

showed that very typical details looked similar in both imaging systems, as can be seen in Fig. 4. These results were significant of course, because they proved that reliable images of animal cells could be made quite easily. The animal tissue did not immediately "burn to a cinder", and the images did not get blurred or distorted by differential loss of electron speed or charging up of the object.

Ruska also paid attention to attempts by Krause to get a better idea of the resolving power of the microscope in Neubabelsberg. His predecessors Driest and Müller thought that they could see details as small as 0.04 μm, which is about five times smaller than what can be achieved with visible light. To get more accurate estimations, Krause had applied a method that is very common in light microscopy. He had used diatom algae, which show repetitive structures consisting of microscopically small ribs. As soon as a repeating pattern can be distinguished, you can count the number of repeated elements per unit of distance. This gives a measure of the smallest thing that can be seen, provided you know the magnification that was used. Krause counted for the species *Amphipleura pellicuda*, for example, 75.2 ribs on a stretch of 10 μm, meaning that the ribs were repeated every 0.133 μm and still distinguishable. In some cases, the distance was even 0.123 μm, according to Krause. This is not as small as Driest and Müller's estimation of the resolution, but it was much more precise proof that Ruska's instrument was capable of surpassing the limits of common light microscopy. Also in the case of this algae species, comparison with light images proved that the electron images showed exactly the same structures. Perhaps even more impressive was the extreme sharpness of the diatom micrographs, because of the very large depth of focus of the electron microscope. As a matter of fact, thanks to this sharpness, it is quite clear that also details can be seen that are much smaller than the repetitive unit itself.

From the same letter that Ruska had sent to Krause in November 1936, it can be learned that Ruska had expected his own lecture to be published in the *Zeitschrift für technische Physik*. This journal had dedicated its two last issues of 1936 to the conference. They did not print Ruska's talk, however, most likely because it contained absolutely no new results of his own. Neither did they publish Krause's results. Krause had already submitted some of his diatom algae images to *Zeitschrift für Physik*—mind the difference in name with *Zeitschrift für technische Physik*—on 21 August, three weeks before the conference, and these were published in October (Krause, 1936). Quite strangely, the paper does not give the result of 0.133 or 0.123 μm as best resolution. The highest resolution here is 0.26 μm, which does not surpass the light microscopical resolution, as Krause admits himself. Neither did the *Zeitschrift für Physik* publish the histological micrographs of Hellbender cells, but this could be due to the physical nature of the journal. Ruska's and Krause's contributions to the conference were finally published in the second half of 1937 in the book *Beiträge zur Elektronenoptik*, edited by Hans Busch and Ernst Brüche (Ruska, 1937; Krause, 1937b). In this book, Krause finally got the opportunity to give a complete presentation of his findings, which includes the very first publication of the Hellbender micrographs. About these images, he asserts that he had obtained comparable micrographs already in the third quarter of 1935, but he had not managed to get them published. It is also the first time that Krause is able to publish his images of the diatom algae with a 0.133 μm grid and remarkable depth of focus. In *Zeitschrift für Physik* as well as the book, Krause explicitly acknowledges August Köhler for his advice and the company Carl Zeiss for supplying a top-quality light microscope.

When it comes to the division between biological transmission electron microscopy and non-biological emission microscopy, it is quite remarkable that at the Bad Salzbrunn conference, Ernst Brüche was finally willing to fully acknowledge its existence. In his own talk, which was soon published in the *Zeitschrift für technische Physik*, he remarked:

"The most general—I might say—classical application of electron optics is the electron microscope. Either it is used to image biological and other objects for the same purpose as in light optics, or to make studies of the emission by cathodes and so

on, as can specifically be done with electron microscopy. In the first case it will be the aim to surpass the resolution limit of the light microscope. Here it were especially the studies at the Technische Hochschule of Berlin, which have brought the problem close to its solution. In the case of electron optical emission studies, which were performed in particular by the AEG Forschungsinstitut, the research goal is knowledge of details of emission processes, which cannot be obtained otherwise."
(Brüche, 1936)

These are historical words. Brüche agreed that there existed a biological transmission electron microscopy in its own right, which was a different application than his emission microscopy, and in one breath he gave the credits for it to the Berlin Polytechnic, of which Ruska was clearly the most important representative by then. As a matter of fact, Max Knoll was also present at the conference, and even gave a lecture, but this was about "Electron optics in television engineering", afterwards published in the same issue of *Zeitschrift für technische Physik* as Brüche's (Knoll, 1936).

Brüche's acknowledgement did not mean, however, that he had drastically changed his opinion on biological transmission electron microscopy. While Ruska ended his talk with voicing the conviction that transmission electron microscopy was on the brink of important developments to come, it was Brüche's opinion that the development of devices analogous to light optical instruments had already reached its peak. Instead, he expected most from innovative applications of electron optics, like television, night vision and electronic multiplier tubes.

4.8 "We Will Be Kings"

It is difficult to tell what the immediate impact was of Ruska's talk at the conference. There were many people in Bad Salzbrunn then, amongst whom was Hermann von Siemens, board member of the huge company with the same name (van Gorkom & de Haas, 1985). It seems reasonable to think that his talk will have had at least some effect on the general opinion about transmission electron microscopy. Whatever the impact may have been, it is a fact that twelve days after his lecture, on 27 September 1936, Ruska received a message from Dr. Harting, board member of Carl Zeiss, saying that he would like to continue the talks that had been broken off 15 months before (Ruska, 1979).

Von Borries and Ruska immediately knew what to do. Actually, from here on it becomes the story of a family business, as Falk Müller called it in 2008 (Müller, 2008). Helmut Ruska had been giving moral support to his brother Ernst for several years, as it has often been told by Ernst Ruska (see for example Ruska, 1970, 1979, 1987), and now the time had come for Helmut to join in actively. Meanwhile Bodo von Borries had been dating Hedwig Ruska for some two years already and was soon going to become their brother-in-law (von Borries, 1991). On 29 September 1936 the three men had a discussion with Helmut's former teacher and current superior, Prof. Dr. Richard Siebeck, director of the *erste Universitätsklinik der Charité*, Berlin's university hospital. Apart from being director of this prestigious hospital, he was a well known heart specialist. He had been a student of the even better known heart specialist Ludolf von Krehl in Heidelberg, with whom Helmut Ruska had studied as well (van Gorkom & de Haas, 1985; Kruger, Schneck, & Gelderblom, 2000). Siebeck felt sympathetic towards their enterprise and was willing to sign a quite outspoken expert opinion, stating for example:

> *"Assuming that an improvement of resolution is only possible to the extent that one can distinguish details ten times smaller than one can see now, and further assuming that one succeeds in developing proper preparation methods—a thing that appears feasible to me—then there is the possibility to solve an abundance of old and new problems in the overall field of medicine. In my opinion, microstructures certainly do not stop at the limit of light microscopical observation." (Ruska, 1979)*

Subsequently a list of potential cell structures is given, that would indeed become important objects of study in future decades. But even more remarkable is a long discourse on the importance of studying bacteria and so-called filter-passing organisms, which we nowadays call viruses. It even contains a reference to the type of filter-passing organisms that would later become known as retroviruses, a class of viruses that may cause certain types of cancer and also AIDS. This remarkable reference is immediately followed by just as noteworthy a reference to bacteriophages, which are viruses that target bacteria and which would very soon become an important research object. Towards the end, Siebeck's expert opinion reads:

> *"Everything that appears to be achievable by now I consider of such relevance, and results are so close as it appears to me, that I am certainly willing to advise on medical research projects and to cooperate by providing my institute's resources." (Ruska, 1979)*

This last statement was very concrete of course, as it meant that Siebeck was really willing to make an effort, and according to Ruska, it did help, as he told in the 1985 interview:

> *"Of course we needed a positive assessment of our business from some other source, otherwise we would not have gotten on with these firms. Siebeck's expert opinion*

was of decisive help in our negotiations. It had another name at the bottom, even when we had written the whole thing ourselves." (van Gorkom & de Haas, 1985)

Siebeck's help was not financial though, as Ruska explained in the interview as well. The support referred to medical research facilities and the issue of necessary medical research licenses. On 2 October 1936, Siebeck put his signature under the text, which had actually been drafted by von Borries and the Ruskas themselves, as we can learn from the last quote. The fact that Siebeck had signed was such a substantial achievement that Ernst Ruska immediately proposed to his girl friend Irmela that very same day (van Gorkom & de Haas, 1985).

Three days later Bodo von Borries spoke with Dr. Carl Köttgen, the Chairman of Siemens Schuckert Werke, and Dr. Heinrich von Buol, Chairman of Siemens & Halske, the sister company of Siemens Schuckert Werke. It turned out that Siemens & Halske would be more suited to house the production of electron microscopes, as it produced all sorts of electrical devices, while Siemens Schuckert was specialised in high voltage engineering. On 10 October von Borries and Ruska went to Jena to speak with the board of Zeiss as well. At this point they actually hoped that a joint venture would be possible with Zeiss and Siemens together, as they argued that Siemens was the best at electrical engineering, while Zeiss was very good at mechanical precision engineering. On 19 October von Borries informed von Buol at Siemens & Halske about the Rüdenberg patent of Siemens Schuckert and their own patents. Finally some of these private patents could strengthen their negotiation position as they had once been intended for in the preceding years. On 23 October they had another talk with Siemens & Halske. At this occasion Dr. Hermann von Siemens was present too, as he was head of the research facilities at the time. On 27 October a further talk followed with Dr. Harting in Jena (Ruska, 1979). After this talk their rally of exactly one month came to a break, as things had to be studied carefully, as it appears.

Bodo von Borries and Ernst Ruska had reason to be much more confident this time. They were talking with two companies now, so it was rather unlikely that the enterprise would fail once more. Meanwhile their friendship had regained its old splendour. Irmela Ruska told about this period:

"I only know that they were together a lot. They were sitting so often together having to discuss so much; I was so jealous then, and I once said to him 'Marry Bodo then!' I just mean to say that they were so intensely engaged in discussions about what to do next, and how to do it. I still remember that at some moment they were

standing on the Kurfürstendamm, and said 'As soon as we make 10,000 Marks a year, we will be kings!'" (van Gorkom & de Haas, 1985)

One could buy four decent middle-class cars for this amount of money.

On 27 November 1936 Ruska and von Borries presented their business plan to Siemens & Halske. This presentation was followed by contract negotiations on 30 November and 16 December. On 18 and 19 December more talks were held with the board of Zeiss Jena as well. Somewhere along the line, however, it had become clear that a joint venture of Siemens and Zeiss was not going to work, as Ruska told in the interview:

"In Germany, Zeiss was the top firm for precision engineering and Siemens was the top firm for electrotechnical engineering. So we thought, maybe they can do it together. But this was discarded quite clearly and annoyed both sides. So, there was competition [apparently]. They did not want to have anything in common." (van Gorkom & de Haas, 1985)

This meant that in the end Ruska and von Borries were confronted with the luxurious problem of having to choose between the two companies. The grounds for their final choice seems to have been a mix of rational and less rational arguments, no doubt involving personal preferences as well. On first sight it appears to be logical that von Borries and Ruska would decide for Siemens & Halske, since Siemens owned the Rüdenberg patent on the electron microscope already, but apparently this patent was not important enough to stop Zeiss from bidding as well. Therefore, the patent cannot have been the only reason. Ruska himself wrote in 1979 that the electrotechnical qualities of Siemens provided the decisive argument (Ruska, 1979). Whatever the case is, on 21 December 1936 the two friends could inform Siemens & Halske that they would like to start on 1 February 1937.

The final agreement between Siemens & Halske and Bodo von Borries and Ernst Ruska to set up a laboratory for the commercial development of electron microscopes was finally signed on 9 January 1937 by the two inventors, as they are referred to in the document, and by Siemens on the 12th, after which it was officially registered on 16 January. Ruska and von Borries agreed to sell a total of 11 patents to the company, for which they would receive 25,000 Reichsmark on 1 February 1937. That is RM 12,500 per person, meaning that by then they had already achieved their goal of becoming "kings", at least for that year.[27] Furthermore they were to receive

[27] Agreement between Siemens & Halske AG and the gentlemen B. von Borries and E. Ruska, SAA 26140 (Siemens Archive). Lin (1995) also quotes from the agreement, but does mention where he found it.

5 percent of the nett selling price for each of the first 100 electron microscopes until 16 March 1950. After the first 100 instruments, the percentage would drop to 3 percent. Since the price of the very first instruments was already RM 60,000, this percentage meant at least an additional bonus of RM 1500 per person per microscope, on top of the income which they would enjoy from their "special employment" by Siemens, as it is called in the agreement.

The date of 16 March 1950 corresponds with the application date of von Borries and Ruska's first two patents, which they had filed in 1932, and which covered the pole-piece lens and the intermediate viewing screen (von Borries & Ruska, 1932b, 1932c). Among the other nine patents was their mutual patent with the application date of 1 December 1933 (von Borries & Ruska, 1933b), and the remainder were the eight individual patents that they filed in December 1934 and in April and May 1935 (von Borries, 1934, 1935; Ruska, 1934c, 1935b). Actually, none of these 11 patents had been granted yet, and for three of them this would never happen.

On closer look, it is not very easy to understand why Siemens finally decided to get involved with the development of electron microscopes. At that moment in late 1936, von Borries and the Ruska brothers had put the main emphasis on biological transmission electron microscopy as follows from the involvement of Helmut Ruska and Siebeck's expert opinion. Nowhere is there any mention of other possible applications, in chemical research or in the steel industry for example. This is slightly surprising, since as a matter of fact these were the industries where the transmission electron microscope was most easily applied as would be proven in the next two years. If it comes to biological and medical applications, however, so far there were only Marton's crude images of plant cells, the micrographs of housefly hairs by Driest and Müller and Krause's micrographs of diatom algae and animal tissue. Marton's first micrograph of bacteria was only to be published in 1937. In retrospect we may say that some of the first biological images looked promising, but at the time it must have been quite a gamble to make such a huge investment in the development of this new technology, as Ruska has stressed himself many years later in his "Early History of Electron lenses and Electron Microscopy" (Ruska, 1979). Nevertheless, Siemens and Halske president Heinrich von Buol appears to have had not so many doubts, writing in July 1943 in a rather unfriendly letter to AEG's Carl Ramsauer:

"Does it need any more evidence that, until the year 1938, when the Siemens Über-mikroskop appeared on the market, AEG did not believe that the Übermikroskop—this typical representative of electron optics—was ripe fruit, ready to pick?" [28]

It is most revealing that von Buol considered the electron microscope to be "ripe fruit" at the time.

5. OTHER COUNTRIES

5.1 Netherlands

One of the few other countries in which serious electron microscopical activities were deployed during the time frame of this chapter, was the Netherlands. On 16 September 1935, the gentlemen Willy G. Burgers and Jan J.A. Ploos van Amstel Esq. had submitted a preliminary report on emission microscopy—called "Cinematographic Record of the $\alpha-\gamma$ Iron Transition as Seen by the Electron-Microscope"—to *Nature*, which was published on 2 November (Burgers & Ploos van Amstel, 1935). The study had been performed in the Physics Laboratory of the electrotechnical company Philips in Eindhoven, which is only a 100 kilometres north east of Brussels, where Marton had just built his third improved transmission microscope. Burgers, the senior of the two researchers, was a physical chemist, just like Marton, but had specialised in diffraction and applied his expertise to the research of solid materials like metals, hence his additional interest in emission microscopy (de Wolff, 1989). As indicated by the two authors, the study was inspired by a paper from 1934 by Ernst Brüche and Walter Knecht. Their microscope, which only needed to render 15- to 25-fold magnifications, contained one magnetic lens, and was based on the microscope design by Knoll, Houtermans, and Schulze (1932a). In November 1936, the two first parts of a comprehensive second report, called "Electron optical observation of metal surfaces" and written in English, were submitted to the Dutch journal *Physica* (Burgers & Ploos van Amstel, 1937)—the same journal in which Marton had just published his theoretical study about surface resolving power versus depth resolving power in transmission microscopy (Marton, 1936c). Two more parts followed in 1938, which were dedicated to zirconium and nickel–iron crystals (Burgers & Ploos van Ams-

---

[28] Letter of Heinrich von Buol to Carl Ramsauer, SAA 11 Flir Lg 177 (Siemens Archive). According to an attached note, the issue was discussed in person by von Buol with AEG-chairman Bücher on 19 July 1943 and a copy of the letter was handed over then. Falk Müller (2009) found the same letter in the archive of the Max Planck Gesellschaft.

tel, 1938). The study by Burgers and Ploos van Amstel is a good illustration of the application of the emission electron microscope first and for all as an instrument. The researchers focussed exclusively on so-called phase transitions of metal surfaces, and the emission microscope itself, which had already fully matured by then, only happened to be a means to that end. Of course, it is illustrative as well that it is an electrotechnical company again, where this emission study took place. Only after World War II would Philips become involved in the production of Jan Bart Le Poole's very successful 'biological' transmission electron microscopes.

5.2 Japan

Eventually, Japan was to become the biggest player in the production of commercial electron microscopes after World War II, with companies like Hitachi, JEOL, Shimadzu and Toshiba. Going by the memoirs of Japanese pioneers, the first two Japanese electron microscopes were of the emission type, both constructed in 1936. The most detailed story is told by Tadatosi Hibi, who built an emission microscope with two magnetic lenses at the Department of Physics of Tohoku University. Just like the Dutch emission microscope, it was based on the design by Knoll, Houtermans and Schulze. It took several years, however, before Hibi managed to obtain proper images with it. According to Hibi's student Keiji Yada, the purpose of Hibi's microscope was to observe phase transitions of metals, just as in the case of the Dutch study (Hibi, 1985; Yada, 1996). According to Koichi Kanaya and Katsumi Ura, the other emission microscope was constructed by Eizi Sugata at Osaka University. It was fitted with only one lens, which was of the magnetic type as well. It appears his interest was a more general understanding of cathode emissions (Kanaya, 1985; Ura, 1996).

5.3 France

It was already mentioned in van Gorkom (2018), that the first French electron microscope appears to have been built by Jean-Jacques Trillat, most likely in 1935. Pierre Grivet, who is considered to be one of the very first French pioneers, wrote in a memoir published in 1985:

> "[Marton's] numerous previous publications in French aroused some academic interest in France but not much action: two experimenters, F. Holweck at the Paris Laboratory of Madame Curie and J.-J. Trillat with his pupil R. Fritz at the University of Besançon, constructed lenses and demonstrated their imaging properties by ob-

taining enlarged images of an oxide cathode; more theoretically minded professors reviewed the basis of electron optics." (Grivet, 1985)

One of these professors was Marton's superior Emile Henriot. One could conclude, therefore, that Grivet tended to include the work from Brussels in the French heritage. With the information currently available, it is impossible to tell whether Fernand Holweck might have been earlier than Trillat and Fritz or not. It seems, though, that it was René Fritz, who published the first proper French article on the electron microscope, as also mentioned in van Gorkom (2018). His article, called "Le Microscope Électronique, Principe, Réalisation et Emploi" (The electron microscope, principles, realisation and application) and published in the *Revue Generale des Sciences pures et appliquées* of 15 June 1936, gives a very straightforward and balanced survey of the theoretical background of the device, its construction and recent achievements, without revealing any preference for either emission or transmission microscopy (Fritz, 1936). Neither did it trigger any French initiatives for several years to come.

6. THE INCUBATION: SUMMARY AND CONCLUSION

A chronological summary of major events will help to draw some conclusions from this chapter and use them to further develop a general overview of the early history of the electron microscope. The summary of this chapter is presented as the following list of 48 most relevant events. Again, it should be emphasised that such a list is rather arbitrary. Its main purpose, however, is to present a timeline, which threads together the events between 1933 and 1937 that took place in separate parts of the world, and which at the same time can serve as an overview.

7 March 1934 Wilfrid Benham from the Canadian Marconi Company presents an emission study to the American Institution of Electrical Engineers. He had built a microscope with one electrical (electrostatic) lens (Benham, 1934).

26 March 1934 Walter Knecht from the AEG Research Institute submits his paper on a hybrid light-, electrical and magnetic emission microscope to *Annalen der Physik* (Knecht, 1934).

9 April 1934 Walter H. Kohl from Rogers Radio Tubes demonstrates an electrical emission microscope during a lecture at the University of Toronto (Hall, 1985).

26–28 April 1934 Chester Calbick and Clinton Davisson from Bell Telephone Laboratories deliver a talk to the American Physical Society on an emission microscope with two electrical lenses (Calbick & Davisson, 1934).

7 May 1934 Ladislaus Marton from the Free University of Brussels submits a publication to *Nature*, called "Electron Microscopy of Biological Objects", which is the first of its kind. It contains the very first histological micrograph and it mentions for the very first time the use of osmium tetroxide as a dye in electron microscopy. It is the first publication on biological electron microscopy in the UK (Marton, 1934a).

8 May 1934 Marton presents another paper, called "Electron Microscopy of Biological Objects, which is published in *Bulletin de la Classe des Sciences de l'Académie Royale de Belgique.*" It mentions for the very first time bacteriology as a field of application (Marton, 1934c).

4 June 1934 Marton presents a third paper on the electron microscopy of biological objects, which is published in *Annales et bulletin de la Société royale des sciences médicales et naturelles de Bruxelles*. It contains the very first micrographs of cell nuclei and the announcement of an electronic shutter (Marton, 1934d).

End of June 1934 Marton visits Berlin to see Ernst Ruska, Ernst Brüche and others.

1 July 1934 Bodo von Borries returns to Berlin in order to intensify his campaign with Ruska.

7 August 1934 Marton submits a fourth publication on the electron microscopy of biological objects, to be published in *Physical Review*. It is the first publication on biological electron microscopy in the USA (Marton, 1934b).

Second half of 1934 Brüche and Scherzer publish *Geometrische Elektronenoptik*—a comprehensive overview of the new discipline, which is rather sceptical about biological transmission microscopy (Brüche & Scherzer, 1934).

29 November 1934 Arnold Sommerfeld and Otto Scherzer from Ludwig-Maximillians-University in Munich publish a critical review of biological electron microscopy in the medical journal *Münchener Medizinische Wochenschrift* (Sommerfeld & Scherzer, 1934).

6 December 1934 Bodo von Borries files a patent on a method to obtain reflection images with an electron microscope (von Borries, 1934).

11 and 13 December 1934 Ernst Ruska files two patents on electron microscope airlocks for objects and photographic plates (Ruska, 1934b, 1934c).

8 January 1935 Arnold Berliner from *Die Naturwissenschaften* does not see the point in publishing the images of housefly hairs by Driest and Müller, which they had obtained at the Berlin Polytechnic.

16 January 1935 Eberhard Driest and Heinz Müller submit their images to *Zeitschrift für wissenschaftliche Mikroskopie und mikroskopische Technik*. They are the first to apply internal photography. They claim a resolution of 0.04 µm (Driest & Müller, 1935).

29 January 1935 Bodo von Borries and Ernst Ruska have a fruitless discussion with Fritz von Wettstein and Georg Melchers at the Kaiser-Wilhelm-Institut für Biologie, failing to gain their support for biological electron microscopy (Ruska, 1979).

Early 1935 Louis Martin from the London Imperial College publishes in *Science Progress* the first British introduction to geometric electron optics in general and more in particular biological transmission microscopy. He is the first to express the desire to visualise viruses (Martin, 1935).

1 February–1 April 1935 Fruitless attempt by Bodo von Borries to raise the interest of IG Farben (Ruska, 1979).

20 February–13 June 1935 A more promising attempt by Ruska and von Borries to interest the Kaiser-Wilhelm-Gesellschaft and Carl Zeiss AG to set up a laboratory for the development of electron microscopes. In the end the talks fail, because of the Rüdenberg patent (Ruska, 1979).

12 March–24 April 1935 Fruitless attempt by von Borries to raise the interest of steel company Krupp (Ruska, 1979).

12 March–4 July 1935 Fruitless attempts by Ruska to raise the interest of Askania, Fernseh AG and Zeiss Ikon for the production of microscopes (Ruska, 1979).

27 April 1935 Bodo von Borries and Ernst Ruska publish an introduction to applied electron optics in *Zeitschrift des Vereines Deutscher Ingenieure*, which they use to promote their ideas about biological electron microscopy (von Borries & Ruska, 1935).

7 May 1935 Marton presents the second of his talks to be published in *Bulletin de la Classe des Sciences de l'Académie Royale de Belgique*." It mentions the very first attempt to use Lenard windows to image living objects (Marton, 1935a).

1 June 1935 Marton presents the third of his talks to be published in *Bulletin de la Classe des Sciences de l'Académie Royale de Belgique*." It is a report on his third magnetic transmission electron microscope, which is the very first with an electronic shutter (Marton, 1935b).

The year 1935 In France, Jean-Jacques Trillat from the Faculty of Sciences of Besançon builds an emission microscope together with René Fritz (Trillat, 1962; Grivet, 1985).

16 September 1935 Willy Burgers and Jan Ploos van Amstel from the Dutch firm Philips publish in *Nature* results obtained with an emission microscope with one magnetic lens (Burgers & Ploos van Amstel, 1935).

Late 1935 Paul Anderson and Kenneth Fitzsimmons from Washington State University construct a high-voltage transmission electron microscope with magnetic lenses and internal photography for biological purposes, but abandon the project in 1938 without publishing results (Reisner, 1989).

Late 1935 Cecil Hall from University of Toronto constructs an emission microscope with one electrical lens (Hall, 1985).

15 November–1 December 1935 Max Knoll publishes an overview of electron microscopy in the medical journal *Zeitschrift für ärztliche Fortbildung* that is more favourable for biological electron microscopy (Knoll, 1935a).

Early 1936 Raphael Liesegang reviews the book *Geometrische Elektronenoptik* in *Zeitschrift für wissenschaftliche Mikroskopie und mikroskopische Technik*. He is particularly sceptical about high-resolution transmission microscopy (Liesegang, 1936).

26 February 1936 Vladimir Zworykin from RCA, New Jersey, gives a talk on electron optical systems at the Berlin Polytechnic, paying specific attention to the hybrid light-electron emission microscope by Walter Knecht from AEG (Zworykin, 1936).

2 April 1936 Reinhold Rüdenberg flees Germany to escape from Nazi prosecution (Rudenberg & Rudenberg, 2010).

May 1936 Marton publishes a paper in *Revue de Microbiologie Appliquée à l'Agriculture, à l'Hygiène et à l'Industrie* which contains the first micrograph ever of cells on a nitrocellulose film (zapon). He claims a resolution of 0.02 μm (Marton, 1936b).

4 June 1936 Otto Scherzer submits to *Zeitschrift für Physik* his classic theory on some errors of electron lenses, which has become known as Scherzer's Theorem (Scherzer, 1936).

15 June 1936 René Fritz from the Faculty of Sciences of Besançon publishes the first French introduction to electron microscopy (Fritz, 1936).

The year 1936 In Japan an emission microscope with two magnetic lenses is built at Tohoku University by Tadatosi Hibi (Hibi, 1985; see also Yada, 1996). At Osaka University Eizi Sugata constructs an emission microscope with one magnetic lens (Kanaya, 1985; Ura, 1996).

8 August 1936 Martin, Whelpton and Parnum submit a paper to *Journal of Scientific Instruments* on a hybrid light-electron transmission microscope with magnetic lenses for biological purposes (Martin et al., 1937).

5 September 1936 Marton submits a theoretical study of surface resolving power versus depth resolving power to *Physica*. The application of nitrocellulose film is mentioned for the first time here (Marton, 1936c).

15 September 1936 Ernst Ruska presents a talk to the 12th German Conference of Physicists and Mathematicians in Bad Salzbrunn, which he uses to further promote biological transmission electron microscopy (Ruska, 1937).

2 October 1936 Richard Siebeck signs an expert opinion, in which there is already the mentioning of studying bacteriophages and carcinogenic viruses (Ruska, 1979).

2 September 1936–16 January 1937 Negotiations by Bodo von Borries and Ernst Ruska with Carl Zeiss and Siemens & Halske, which finally materialise in an agreement with Siemens & Halske (Ruska, 1979).

5 December 1936 Marton presents the fourth of his talks to be published in *Bulletin de la Classe des Sciences de l'Académie Royale de Belgique*. Here he claims a resolution of 0.01 μm (Marton, 1936a).

Late 1936 Cecil Hall constructs a simple transmission microscope with two magnetic lenses for biological purposes (Hall, 1985).

1 February 1937 Beginning of the development of transmission electron microscopes at Siemens & Halske by Bodo von Borries and Ernst Ruska.

5 April 1937 Howard McMillen and Gordon Scott from Washington University in St. Louis submit a paper on a magnetic emission microscope with one lens for biological purposes to *Review of Scientific Instruments* (McMillen & Scott, 1937).

5 June 1937 Marton presents the fifth of his talks to be published in *Bulletin de la Classe des Sciences de l'Académie Royale de Belgique*. It contains the first micrographs of bacteria ever (Marton, 1937).

Second half of 1937 Friedrich Krause's contribution to the 12th German Conference in Bad Salzbrunn appears in the book *Beiträge zur Elektronenoptik*. Krause claims a resolution of 0.123 μm (Krause, 1937b).

Assuming that indeed this list is a proper representation of the most relevant events during the years 1934 till 1937, it will be helpful to break it down in rough numbers in order to gain a better perspective. The resulting statistics cannot be anything more than crude, however, since the grouping together of certain closely-related events is evidently rather arbitrary.

6.1 Spread Across the Globe

Already from the structure of this chapter it is clear that electron microscopy finally starts to spread on an international scale in this period. For the 53 entries that were listed in total for the period 1928 till 1933 in van Gorkom (2018), only three refer to work being done outside Germany: these were contributions by Davisson and Calbick, Zworykin and Marton, which is roughly 5 percent. In the period 1934 till 1937 we see the number of entries from outside Germany increase to 25 out of 48, which is roughly 50 percent. Biggest contributor to this development is Marton in Belgium with 11 entries, followed by the USA and Canada with 4 each.

Effectively, this international increase implies a decrease in Germany itself, since the average total of major events is practically the same for 1933, 1934, 1935 and 1936. It means that German activity is halved in 1934, 1935 and 1936, and the decline would have been much stronger if all campaigning by Ruska and von Borries had been excluded from the list. To phrase it differently, not much practical work was done in Germany any more. Substantial German results were limited to Knecht's hybrid microscope in 1934, the images of housefly hairs by Driest and Müller in 1935, and in 1936 Scherzer's Theorem and Krause's images of diatom algae. I think, we may safely conclude therefore that it was especially Marton who kept the fire of electron microscopy burning. At least, he should be acknowledged for having been the first to focus emphatically on the "electron microscopy of biological objects," as his first four papers on the subject were called, the first to publish on this subject in the UK and the USA, the first to image cells and cell nuclei, the first to apply osmium tetroxide as an electron microscopical dye, the first to pay an international visit to likewise-minded colleagues, the first to try to image living objects, the first to use a nitro-cellulose film as an object carrier, the first to mention the wish to apply the instrument to bacteriology and also the first to actually publish bacteriological micrographs.

6.2 Magnetic Transmission School

This major contribution to the development of biological electron microscopy brings us to the rise of two different electron microscopical schools as I already discussed in the summary and conclusions of van Gorkom (2018). One of them is what I have called the magnetic transmission school. Probably the most important trait that the members of this school share, is their primary interest in applying the nascent technology to biological and medical issues in essentially the same way as it is done in light microscopy. This automatically implies transmission microscopy, and since you need high tensions for large magnifications, it also implies the choice for magnetic lenses, since electrical lenses will very easily short-circuit if applied that way. It cannot be said, however, whether the earliest members of the school were really fully aware of this. Another explanation of the choice for magnetic lenses is a far more trivial one, being the fact that Ruska had set the example with the polepiece lens. Another important trait of this school is that basically all its members belonged to a university, when they made their contributions. This should be explained from the fact that it was very hard to imagine any commercial value of transmission electron microscopy, and therefore industry would not participate in its development, at least, not until the end of 1936.

No doubt, it is Marton who positions himself in this period as the school's most prominent representative. He is very active throughout the whole period—on an international scale even, and fits all the criteria, when developing biological high-tension magnetic transmission microscopy at a university. Rather ironically, this is far less the case for Ernst Ruska and Bodo von Borries who are occupied as common electrotechnical engineers in the television and high-voltage industry respectively, and are not able to make any tangible contributions to the new field. Nevertheless, in reality they are probably the most active advocates of magnetic transmission microscopy in their efforts to initiate an industrial development, while they earned their credits in an academic environment as well.

Apart from these three men, I have also presented Reinhold Rüdenberg, Léo Szilárd and Paul Anderson as members of this school. Rüdenberg could be seen as atypical, since he worked for the Siemens-Schuckert

Werke at the time, but on the other hand he was also professor at Berlin's Technische Hochschule then. Moreover, his contribution did not involve any actual development or experiment, but merely an idea in the form of a patent, giving him a rather atypical position anyway. The same may be said for Szilárd, although he meets the academic criteria fully. Paul Anderson, finally, did not undertake any development himself, but he was soon going to be joined by Kenneth Fitzsimmons, who did the practical work, which fits all the criteria for the transmission school.

In this chapter, the school is subsequently expanded with Eberhard Driest, Heinz Müller and Friedrich Krause in Germany, Louis Martin and collaborators in England, and Eli Burton and Cecil Hall in Canada. All of them fit the criteria with a slight exception for Hall, since his first trial was an electrical emission microscope before he devoted himself to a true transmission microscope with magnetic lenses. It appears to be justified to add Gordon Scott and Howard McMillen from Washington University in St. Louis as well, although their case is a bit more complicated. Scott was a pathologist, who was interested in the emission by minerals contained in biological substances, and therefore one could argue that this was an emission study as well. Nevertheless, they do meet all the other criteria of the magnetic transmission school and Scott eventually turned to transmission microscopy as well, making him a full-grown member of this school in the end. Altogether, if we count in 'operational units' instead of individuals, we see an increase from five to ten units in the period 1934 till 1937. These ten units are Ernst Ruska with von Borries and Helmut Ruska, Rüdenberg, Szilárd, Anderson with Fitzsimmons, Marton, Müller, Krause, Martin with Whelpton and Parnum, Burton with Hall, and Scott with McMillen.

Probably the most articulate new member of the school is Louis Martin, who made a very clear distinction between "two classes of instruments" in the paper in which he and his colleagues presented their hybrid light-electron microscope, referring to the emission microscope at the one hand and the transmission microscope "for practical microscopy" at the other (Martin et al., 1937). Burton was just as outspoken, if we go by Cecil Hall (Hall, 1985), but during this period he does not publish on the subject yet. Of course, the school's most vocal representative in Germany was Ernst Ruska who used his talk at the 1936 German Conference of Physicists and Mathematicians in Bad Salzbrunn to question whether it really made sense to call emission studies "microscopy" (Ruska, 1937).

At this stage, it also becomes clear that Ruska should be seen as the most authoritative member of this school. All the work of the units Anderson, Marton, Müller, Krause, Martin and Burton is based on papers by Ruska, either published with Knoll, or alone. Nobody mentions von Borries, who had published only two research papers with Ruska, which had less significance to the field (von Borries & Ruska, 1932a, 1933a). Evidently, the relation between Ruska's contributions and those of Rüdenberg and Szilárd is completely different, as was already discussed extensively in van Gorkom (2018). But also if we acknowledge the legitimacy of the Rüdenberg–Siemens patents, it can still be said that Ruska has been a member of the school the longest, since his construction of the first prototype of a transmission electron microscope predates the patents. Moreover, Ruska remained active in the field, while Rüdenberg did not engage in any activity ever since filing the Siemens patents. To Szilárd, more or less identical arguments can be applied, especially since the report on his earliest suggestion of a transmission microscopy is based on a single anecdote by one man about an incidental remark, made in passing.

6.3 Electric Emission School

The school that opposed the magnetic transmission school, I have called the electric emission school. It refers to the research of cathode emissions by means of cathode-ray tubes, which have often only one electric lens. For experienced researchers of cathode-ray tubes, it was rather easy to adjust the device in such a way that it would produce enlarged images of the cathode, and soon some spectacular results were obtained this way, of which the AEG-film of cathode emissions, shown on 17 June 1932 at the meeting of the German Physical Society and the Society of Applied Physics in Berlin, appears to have been the absolute highlight (Brüche & Johannson, 1932). Since results can be obtained easily and have immediate relevance to the manufacturing of oscilloscopes, televisions and the like, emission microscopy is especially interesting to electrotechnical companies, where we find practically all members of this school. It should be noted, however, that the commitment to electrical lenses of this school is less outspoken than the transmission school's preference for magnetic lenses. A tendency to apply magnetic lenses is started by Knoll, Houtermans and Schulze who used Knoll's 'own' encased magnetic lens for their emission studies. This tendency should be explained by the fact that the applied magnetic lenses were fitted at the outside of the tube, giving the possibility to shift them, which made focussing far easier (Knoll, Houtermans, & Schulze, 1932b).

Ernst Brüche of the AEG Research Institute is certainly the most vocal member of the emission school, if we go by his vast editorial activities, like the publishing of the book *Geometrische Elektronenoptik*, together with Otto Scherzer, and the book *Beiträge zur Elektronenoptik*, together with Hans Busch. I count as members of his operational unit Helmut Johannson and Otto Scherzer, although Scherzer moved to the Ludwig Maximillians University of Munich later on. In van Gorkom (2018), I also presented Max Knoll and Vladimir Zworykin as members of this school, because of their preferences for emission studies. Zworykin meets all criteria for the electrical emission school as he is a representative of RCA. Knoll meets the commercial criteria as employee of Telefunken.

In this chapter, we saw that Walter Knecht built a hybrid light–electron microscope for emission studies. As an employee of AEG, he clearly belongs to Brüche's unit. Furthermore, there were Chester Calbick and Clinton Davisson who presented a paper on an electrical emission microscope, which qualifies them as members of the electrical school as well, especially since they are employed by Bell Telephone Laboratories. Furthermore, we can add Wilfrid Benham from the Canadian Marconi Company, Walter Kohl from the Canadian company Rogers Radio Tubes and Willy Burgers and Jan Ploos van Amstel from Philips in the Netherlands. The Dutch case is an interesting one, since it is an outspoken example of the application of the emission microscope as an instrument in its own right. At the same time, they are also representatives of the Knoll faction as they employ a magnetic lens instead of an electrical one. When counting 'operational units' again, we notice an increase for the emission school as well. We started with three: Brüche with Johannson, Scherzer and Knecht, Zworykin, and Knoll with Houtermans and Schulze. This number increases to seven after adding the units Calbick with Davisson, Benham, Kohl, and Burgers with Ploos van Amstel.

It would certainly be fair to name Ernst Brüche also as the most senior member of the emission school. One could counter this claim with the notion that Max Knoll had already patented an electron lens in 1929, while Brüche published his first contribution about electron optics in 1930, but then it should be stressed that Brüche had always been present at the forefront ever since he entered the field and had always been very adamant about his role, as illustrated by his talk at the 1936 German Conference of Physicists and Mathematicians in Bad Salzbrunn, where Ruska questioned the legitimacy of the emission school. In contrast to this, there is no evidence at all that Knoll regarded himself as a representative of his school and he even tended to side with Ruska as can be learned from the introduction to electron microscopy, which he wrote for a medical journal in

1935 (Knoll, 1935a). In fact, the outspoken rivalry between Ernst Ruska and Ernst Brüche could very well be used as another argument to picture them as the leaders of the two schools. This rivalry will only intensify.

As could be expected, we are left with a few cases of pioneering work which cannot be fitted into one of the two schools so easily. This applies to the developments in France and Japan. In France two emission microscopes were built by Fernand Holweck and Jean-Jacques Trillat. In Japan two emission microscopes were built by Tadatosi Hibi and Eizi Sugata. All four scientists were working at academic institutes, and therefore they do not immediately qualify as members of the emission school.

The two French experiments are hardly documented and received no follow-up at all. They seem to have come purely from a theoretical physical interest in geometric electron optics. The two Japanese experiment with magnetic emission microscopes are neither well documented. It appears that Hibi's work was very similar to the work done by Burgers and Ploos van Amstel at Philips. Sugata appears to have had a more general interest in cathode emissions. For these reasons, it would be justified to label these four researchers as late and atypical members of the emission school. However, especially in the case of the two French experiments, there are just as good reasons not to label them at all and accept the fact that you cannot categorise everybody.

6.4 Dominance of the Transmission School

From here we can return to the listing of 48 events with which this final section started, to perform some very simple statistics again. However, before I do so, I like to stress once more that the numbers that can be derived from the list are certainly arbitrary. Having said this, it turns out that 26 entries on the list refer to actual research. About two thirds of these 'research' entries (17) are linked to the transmission school, and about one third to emission (9), including the French and Japanese work. If we disregard the French and Japanese entries, the number of major emission-related events goes down to 7, which is about a quarter. From these numbers can be concluded, that the major events between 1933 and 1937 are dominated by transmission work. This is in line with the view that the emission school already peaked in 1932 and had little new to offer after that.

The other 22 entries on the list relate in one way or another to what you might call school-building—with the possible exception of three entries. More than half of these 22 entries (13) concern the efforts of Ruska and von Borries to get the commercial development of transmission microscopes going. Apart from these 13 events, 2 more could be considered beneficial to the transmission case, meaning that again two-third of the entries is transmission school related (15 out of 22). From the remaining 7 'school-building' entries only 4 are clearly beneficial to the emission school. The remainder of 3 you might call neutral events, which are Rüdenberg's escape, the publication of Scherzer's Theorem and the French introduction by Fritz. To put it shortly: whatever way you look at it, the transmission school is at least twice as active as the emission school during this period. This activity of the transmission school certainly fits the picture of a nascent discipline which is struggling for recognition.

The critical reader might make the understandable objection that my conclusion is biased by an apparent overemphasis on the activities of the transmission school. I like to counter this objection with the observation that 17 out of 33 events, listed in the previous section were related to the emission school. It means, that even if there exists a bias, one can still see a decline in activity from approximately 50 percent to approximately 33 percent.

Whatever way you look at the observed balance between the two schools, at least I should have provided sufficient arguments that indeed two electron microscopical schools existed between 1931 and 1937. As already mentioned at the conclusion of van Gorkom (2018), this takes us to Thomas Kuhn's thoughts on schools as explained by him in his classic essay "The Structure of Scientific Revolutions" (Kuhn, 1962).

6.5 Motives

The question of motives is only intensified by the emergence of the two schools, which I have painted here. After all, apart from the determination to develop electron microscopes as such, it also turns out that a preference existed for one or the other approach.

For the electrical school the answer appears to be very straightforward, as has been pointed out before already. For electrotechnical companies, it was rather easy to construct an emission microscope and the results of cathode studies had immediate relevance to other commercial developments. The ease and speed with which these results were accomplished also explain why emission microscopy was to die down again, since the actual applications are very limited.

More complex is the situation for magnetic transmission microscopy, which is first and for all characterised by its emphasis on biological and medical applications. The very first remarks on such applications were made by Ernst Ruska in two of his final papers as active developer, which he submitted by the end of 1933 (Ruska, 1934a, 1934d). He was immediately succeeded by Ladislaus Marton in Belgium who made the biological application of the instrument his central theme. Then the Germans Ernst Driest and Heinz Müller came with their house fly and finally there was Friedrich Krause with his diatom algae and hellbender cells. Within the time frame of this chapter, outspoken verbal support for biological applications was supplied especially by Louis Martin in the UK and Eli Burton in Canada. What strikes one is the fact that we find only few medical researchers among the school members, and no biologists at all. Ruska and von Borries are electrotechnical engineers, Rüdenberg too, Marton is a physical chemist, and Szilárd, Anderson, Martin and Burton are physicists. Only in the background of Ernst Ruska we find Helmut Ruska, who is a medical student at first, and there is Friedrich Krause, who is still a medical student as well. Therefore, the only remaining fully qualified medical researcher is the pathologist Gordon Scott. Originally, however, Scott's primary interest was an emission study of minerals contained by histological objects, which were mounted on the cathode, so it seems that originally Scott neither shared the belief in the possibility to apply transmission microscopy to biological and medical studies.

At the same time, this absence of solid support from biologists and medical researchers is accompanied by rather substantial criticism from many sides. In van Gorkom (2018) and Section 1, we already met the bacteriologist Jules Bordet, who was not charmed by the prospect of a new type of microscope. In this chapter, we saw that the physicists Ernst Brüche and Otto Scherzer argued that you cannot image a molecule, since any electron hitting it will increase its momentum dramatically. The underlying assumption was that the submicroscopical world would be completely fluid, only containing free moving molecules and atoms, while image formation would depend on electron absorption. Scherzer and the renowned physicist Arnold Sommerfeld added that absorption will cause differences in electron velocities and subsequently chromatic aberration, objects will be damaged, and they will charge up. Biologists Fritz von Wettstein and Georg Melchers supported the concept that fixed structures do not exist at a submicroscopical level. The well-known colloidal chemist Raphael Liesegang shared this belief as well. And finally Gustav Hertz is also reported to have been critical of electron microscopy (Lin, 1995).

Apart from this we saw other significant lack of interest. Chief-editor of *Die Naturwissenschaften* Arnold Berliner simply refused to publish the images of Eberhard Driest and Heinz Müller, while Friedrich Krause did not manage to publish his results at all for more than a year. And finally Bodo von Borries was not able to convince industries like IG Farben and Krupp, despite the fact that obtaining results could be expected to be easier in colloidal and metallurgic research for the reason that inorganic materials are less vulnerable.

In fact, the only substantial external support for transmission microscopy came from Max Planck, which helped to keep the negotiations with Carl Zeiss going for some time, and finally from the medical specialist Richard Siebeck, whose cooperation was decisive for the final agreement with Siemens. It should immediately be noted once more, however, that Siebeck's role was limited to signing an expert opinion, which had actually been written by the Ruska brothers and von Borries themselves. So, in the end it was only thanks to the sudden view of Siemens that the electron microscope was "ripe fruit, ready to pick"—as Heinrich von Buol called it in 1943 in his letter to AEG—the Ruska and von Borries managed to succeed in their effort to get the industrial production of electron microscopes going.

All this does not answer the question of what motivated men like Ruska, von Borries, Marton and Krause to pursue a goal which was deemed hopeless on rather sound grounds by people who ought to know. If it comes to the true academics among the school members, like Paul Anderson, Eli Burton, Gordon Scott and Louis Martin, it can be certainly accounted to the mere scientific curiosity, whether it would be possible to make something like a transmission electron microscope work. In the case of the other men, however, we see a remarkable perseverance, especially with Ruska and von Borries who seriously invested lots of time and money in a possible future of the instrument. In the case of the latter two we may even say that their dedication bordered on irrationality—a view which was certainly shared by Ernst Ruska himself. When in 1985 we tried to find out what had motivated him, he exclaimed:

> *"Yes, it was a mania! We had sunk our teeth in the thought of it." (van Gorkom & de Haas, 1985)*

And two days later:

> *"It is certainly true that we had closed our eyes for many things, otherwise we would not have had the 'push' to go on with it. That is irrational. It is a kind of monomania.*

That was also there." (Ruska used the English word "push"; van Gorkom & de Haas, 1985)

Probably, their mania can be interpreted best as the obsessive dedication of any person who is determined to set up some sort of business of his own. In their case, they ended up selling their assets to Siemens & Halske, but it brought them finally the wealth "to be kings" and the chance to develop their own magnetic transmission electron microscope without having to carry the burden of huge investments themselves.

REFERENCES

Baker, J. R. (1958). *Principles of Biological Microtechnique; A Study of Fixation and Dyeing.* London: Methuen.

Beneke, K. (2004/2006). *Liesegang named in literature (2004, 2006).* http://www.uni-kiel.de/anorg/lagaly/group/klausSchiver/liesegangliterature-1.pdf. (Accessed 3 August 2013).

Benham, W. E. (1934). Note on a demonstration of a low-voltage electron microscope using electrostatic focusing. *Journal of the Institution of Electrical Engineers, 75*, 388–390.

Bethe, A. (1903). *Allgemeine Anatomie und Physiologie des Nervensystems.* Leipzig: Thieme.

Brüche, E. (1936). Übersicht über die experimentelle Elektronenoptik und ihre Anwendung. *Zeitschrift für technische Physik, 17*, 588–593.

Brüche, E., & Johannson, H. (1932). Kinematographische Elektronenmikroskopie von Oxydkathoden. *Annalen der Physik, 407*, 145–166.

Brüche, E., & Scherzer, O. (1934). *Geometrische Elektronenoptik.* Berlin: Springer.

Burgers, W. G., & Ploos van Amstel, J. J. A. (1935). Cinematographic record of the α–γ iron transition as seen by the electron–microscope. *Nature, 136*, 721.

Burgers, W. G., & Ploos van Amstel, J. J. A. (1937). Electronoptical observation of metal surfaces I and II. *Physica, 4*, 5–14, 15–22.

Burgers, W. G., & Ploos van Amstel, J. J. A. (1938). Electronoptical observation of metal surfaces III and IV. *Physica, 5*, 305–312, 313–319.

Burton, E. F. (1948). Obituary. *Physics Today, 1*, 28.

Calbick, C. J., & Davisson, C. J. (1934). Electron microscope. *Physical Review, 45*, 764.

Davisson, C. J., & Calbick, C. J. (1931). Electron lenses. *Physical Review, 38*, 585.

Davisson, C. J., & Calbick, C. J. (1932). Electron lenses (Corrigendum). *Physical Review, 42*, 580

de Wolff, P. M. (1989). Levensbericht W.G. Burgers. In *Jaarboek 1989* (pp. 134–140). Amsterdam: Huygens Institute–Royal Netherlands Academy of Arts and Sciences (KNAW).

Driest, E., & Müller, H. (1935). Elektronenmikroskopische Aufnahmen (Elektronenmikrogramme) von Chitinobjekten. *Zeitschrift für wissenschaftliche Mikroskopie und mikroskopische Technik, 52*, 53–57.

Forman, P., & Hermanns, A. (2008). Sommerfeld, Arnold (Johannes Wilhelm). In *Complete Dictionary of Scientific Biography: Vol. 12.* Detroit: Charles Scribner's Sons (pp. 525–532). Gale Virtual Reference Library, http://go.galegroup.com/ps/. (Accessed 3 August 2013).

Fritz, R. (1936). Le microscope électronique. *Revue Generale des Sciences Pures et Appliquées, 47*, 338–342.

Gabor, D. (1968). Preface. In L. Marton, *Early History of Electron Microscopy*. San Francisco: San Francisco Press.

Golgi, C. (1898). Intorno alla struttura delle cellule nervose. *Bollettino della Società Medico-Chirurgica di Pavia, 13*, 3–16.

Golgi, C. (1967). The neuron doctrine: Theory and facts. Nobel lecture, 11 December 1906. In *Nobel Lectures, Physiology or Medicine 1901–1921* (pp. 189–217). Amsterdam: Elsevier Publishing Company. The Golgi apparatus can be found in practically all eukaryotic cells. Golgi received in 1906 the Nobel Prize in Physiology or Medicine together with Santiago Ramón y Cajal "in recognition of their work on the structure of the nervous system". www.nobelprize.org.

Grivet, P. (1985). The French electrostatic electron microscope (1941–1952). In P. W. Hawkes (Ed.), *Advances in Electronics and Electron Physics: Supplement 16. The beginnings of electron microscopy* (pp. 225–274). Orlando: Academic Press.

Gütt, A. (1935). Das Ehetauglichkeitszeugnis, Gesetz zum Schutz der Erbgesundheit des deutschen Volkes. *Zeitschrift für ärztliche Fortbildung, 32*, 663–664.

Hall, C. E. (1985). Recollections from the early years: Canada–USA. In P. W. Hawkes (Ed.), *Advances in Electronics and Electron Physics: Supplement 16. The beginnings of electron microscopy* (pp. 275–296). Orlando: Academic Press.

Hibi, T. (1985). My recollection of the early history of our work on electron optics and the electron microscope. In P. W. Hawkes (Ed.), *Advances in Electronics and Electron Physics: Supplement 16. The beginnings of electron microscopy* (pp. 297–315). Orlando: Academic Press.

Hickethier, K. (2008). Early TV: Imagining and realizing television. In J. Bignell, & A. Fickers (Eds.), *A European Television History*. London: Wiley-Blackwell.

Johannson, H. (1933). Über das Immersionsobjektiv der geometrischen Elektronenoptik. *Annalen der Physik, 410*, 385–413.

Johannson, H., & Scherzer, O. (1933). Über die elektrische Elektronensammellinse. *Zeitschrift für Physik, 80*, 183–192.

Kanaya, K. (1985). Reminiscences of the development of electron optics and electron microscope instrumentation in Japan. In P. W. Hawkes (Ed.), *Advances in Electronics and Electron Physics: Supplement 16. The beginnings of electron microscopy* (pp. 317–382). Orlando: Academic Press.

Knecht, W. (1934). Das kombinierte Licht- und Elektronenmikroskop, seine Eigenschaften und seine Anwendung. *Annalen der Physik, 412*, 161–182.

Knoll, M. (1935a). Das Elektronenmikroskop. *Zeitschrift für ärztliche Fortbildung, 32*, 644–647, 678–679.

Knoll, M. (1935b). Aufladepotential und Sekundäremission elektronenbestrahlter Körper. *Zeitschrift für technische Physik, 16*, 467–475.

Knoll, M. (1936). Die Elektronenoptik in der Fernsehtechnik. *Zeitschrift für technische Physik, 17*, 604–617.

Knoll, M., Houtermans, F. G., & Schulze, W. (1932a). Untersuchung der Emissionsverteilung an Glühkathoden mit dem magnetischen Elektronenmikroskop. *Zeitschrift für Physik, 78*, 340–362.

Knoll, M., Houtermans, F. G., & Schulze, W. (1932b). Über geometrisch-optische Abbildung von Glühkathoden durch Elektronenstrahlen mit Hilfe von Magnetfeldern (Elektronenmikroskop). *Verhandlungen der deutschen physikalischen Gesellschaft, 13*, 23–24.

Knoll, M., Houtermans, F., & Schulze, W. (1932c). Elektronenmikroskop, bei dem Elektronen aussendende Substanzen in vergrößertem Maßstabe abgebildet werden. German patent 679330, priority date of 17 March 1932, published on 13 July 1939.

Knoll, M., Houtermans, F., & Schulze, W. (1934). Electron microscope. US patent 2,131,536, priority date of 23 June 1934, granted on 27 September 1938.

Knoll, M., & Ruska, E. (1932a). Beitrag zur geometrischen Elektronenoptik (I–II). *Annalen der Physik*, *404*, 607–661.

Knoll, M., & Ruska, E. (1932b). Das Elektronenmikroskop. *Zeitschrift für Physik*, *78*, 318–339.

Krause, F. (1936). Elektronenmikroskopische Aufnahmen von Diatomeen mit dem magnetischen Elektronenmikroskop. *Zeitschrift für Physik*, *102*, 417–422.

Krause, F. (1937a). Das magnetische Elektronenmikroskop und seine Anwendung in der Biologie. *Naturwissenschaften*, *25*, 817–825.

Krause, F. (1937b). Neuere Untersuchungen mit dem magnetischen Elektronenmikroskop. In H. Busch, & E. Brüche (Eds.), *Beiträge zur Elektronenoptik* (pp. 55–61). Leipzig: Johann Ambrosius Barth.

Kruger, D. H., Schneck, P., & Gelderblom, H. R. (2000). Helmut Ruska and the visualisation of viruses. *The Lancet*, *355*, 1713–1717.

Kuhn, T. S. (1962). *The structure of scientific revolutions*. Chicago and London: The University of Chicago Press. Third edition, 1996.

Lambert, L., & Mulvey, T. (1996). Ernst Ruska (1906–1988), designer extraordinaire of the electron microscope: A memoir. *Advances in Imaging and Electron Physics*, *95*, 3–62.

Liesegang, R. E. (1936). Rezension zu Brüche und Scherzer 1934. *Zeitschrift für wissenschaftliche Mikroskopie und mikroskopische Technik*, *53*, 227.

Lin, Q. (1995). *Zur Frühgeschichte des Elektronenmikroskops*. Stuttgart: Verlag für Geschichte der Naturwissenschaften und der Technik.

Martin, L. C. (1935). Electron optics. *Science Progress*, *29*, 426–437.

Martin, L. C., Whelpton, R. V., & Parnum, D. H. (1937). A new electron microscope. *Journal of Scientific Instruments*, *14*, 14–24.

Marton, L. (1934a). Electron microscopy of biological objects. *Nature*, *133*, 911.

Marton, L. (1934b). Electron microscopy of biological objects. *Physical Review*, *46*, 527–528.

Marton, L. (1934c). La microscopie électronique des objets biologiques. *Bulletin de la Classe des Sciences de l'Académie Royale de Belgique*, *20*, 439–446.

Marton, L. (1934d). Le microscope électronique, premiers essais d'application à la biologie. *Annales et Bulletin de la Société Royale des Sciences Médicales et Naturelles de Bruxelles*, 92–106.

Marton, L. (1934e). Le microscope électronique et ses applications. *Bulletin de la Société Française de Physique*. No. 364.

Marton, L. (1935a). La microscopie électronique des objets biologiques II. *Bulletin de la Classe des Sciences de l'Académie Royale de Belgique*, *21*, 553–564.

Marton, L. (1935b). La microscopie électronique des objets biologiques III. *Bulletin de la Classe des Sciences de l'Académie Royale de Belgique*, *21*, 606–617.

Marton, L. (1935c). Le microscope électronique et ses applications. *Revue d'Optique, Théorique et Instrumentale*, *14*, 129–145.

Marton, L. (1936a). La microscopie électronique des objets biologiques IV. *Bulletin de la Classe des Sciences de l'Academie Royale de Belgique*, *22*, 1336–1344.

Marton, L. (1936b). Le microscope électronique. *Revue de Microbiologie Appliquée à l'Agriculture, à l'Hygiène et à l'Industrie*, *11*, 117–124.

Marton, L. (1936c). Quelques considérations concernant le pouvoir séparateur en microscopie électronique. *Physica*, *3*, 959–967.

Marton, L. (1937). La microscopie électronique des objets biologiques V. *Bulletin de la Classe des Sciences de l'Academie Royale de Belgique, 23*, 672–675.

Marton, L. (1968). *Early History of Electron Microscopy*. San Francisco: San Francisco Press.

Marton, L. (1976). Early application of electron microscopy to biology. *Ultramicroscopy, 1*, 281–296.

Marton, L., & Nuyens, M. (1933). Meetkundige optiek der electronen. *Wis- en Natuurkundig Tijdschrift, 6*, 159–170.

McMillen, J. H., & Scott, G. H. (1937). A magnetic electron microscope of simple design. *Review of Scientific Instruments, 8*, 288–290.

Müller, F. (2008). A scientific-technological system as a family business: The Ruskas, Bodo von Borries and the electron microscope. In A. Presas i Puig (Ed.), *Who is making science? Scientists as makers of technical-scientific structures and administrators of science policy* (pp. 21–33). MPI-Preprint 361, Berlin.

Müller, F. (2009). The birth of a modern instrument and its development during World War II: Electron microscopy in Germany from the 1930s to 1945. In A. Maas, & H. Hooijmaijers (Eds.), *Scientific Research in World War II, what Scientists did in the War* (pp. 121–146). Abingdon: Routledge.

Mulvey, T. (1985). The industrial development of the electron microscope by the Metropolitan-Vickers Electrical Company and AEI Limited. In P. W. Hawkes (Ed.), *Advances in Electronics and Electron Physics: Supplement 16. The beginnings of electron microscopy* (pp. 417–442). Orlando: Academic Press.

Nipkow, P. (1884). Elektrisches Teleskop. German patent 30105, priority date of 6 January 1884, published on 15 January 1885.

Olby, R. (1994). *The path to the double helix, the discovery of DNA*. New York: Dover Publications. Staudinger received in 1953 the Nobel Prize in Chemistry "for his discoveries in the field of macromolecular chemistry". www.nobelprize.org.

Purkayastha, M., & Williams, R. P. (1960). Association of pigment with the cell envelope of *Serratia marcescens* (*Chromobacterium prodigiosum*). *Nature, 187*, 349–350.

Reisner, J. H. (1989). An early history of the electron microscope in the United States. *Advances in Electronics and Electron Physics, 73*, 133–231.

Rosing, B. (1907). Verfahren zur elektrischen Fernübertragung von Bildern. German patent 209320, priority date of 26 November 1907, granted on 24 April 1909.

Rüdenberg, R. (1932). Einrichtung zum vergrößerten Abbilden von Gegenständen durch Elektronenstrahlen (Elektronenmikroskop). German patent 916 841, priority date of 31 March 1932, granted 8 July 1954.

Rudenberg, G., & Rudenberg, P. G. (2010). Origin and background of the invention of the electron microscope: Commentary and expanded notes on memoir of Reinhold Rüdenberg. In P. Hawkes (Ed.), *Advances in Imaging and Electron Physics: Vol. 160*. Academic Press (pp. 207–286).

Ruska, E. (1930). *Untersuchung elektrostatischer Sammelvorrichtungen als Ersatz der magnetischen Konzentrierspulen beim Kathodenstrahloszillographen* (Diplomarbeit). Technische Hochschule Berlin, Lehrstuhl für Hochspannungstechnik.

Ruska, E. (1934a). Über Fortschritte im Bau und in der Leistung des magnetischen Elektronenmikroskops. *Zeitschrift für Physik, 87*, 580–602.

Ruska, E. (1934b). Einschleusvorrichtung für an der Pumpe betriebene Korpuskularstrahlapparate. German patent 659092, filed on 12 December 1934, granted on 31 March 1938.

Ruska, E. (1934c). Vorrichtung zum Einbringen von einzuschleusenden Gegenständen in das Vakuum von Korpuskularstrahlen. German patent 874819, filed on 14 December 1934, published on 20 November 1941, and granted on 12 March 1953.

Ruska, E. (1934d). Das Elektronenmikroskop als Übermikroskop. *Forschungen und Fortschritte, 10*, 8.

Ruska, E. (1935a). The electron microscope as ultra-microscope. *Research and Progress, 1*, 18–19.

Ruska, E. (1935b). Kreisringförmige Polschuhe für magnetische Elektronenlinsen. German patent 730719, priority date of 27 April 1935, granted on 17 December 1942.

Ruska, E. (1937). Elektronenmikroskop und Übermikroskop. In *Beiträge zur Elektronenoptik* (pp. 49–54). Leipzig: Johann Ambrosius Barth.

Ruska, E. (1970). Erinnerungen an die Anfänge der Elektronenmikroskopie. In *Festschrift anlässlich der Verleihung des Paul-Ehrlich- und Ludwig-Darmstaedter-Preises 1970*. Stuttgart: Gustav-Fischer-Verlag.

Ruska, E. (1974). Zur Vor- und Frühgeschichte des Elektronenmikroskops. In *Electron microscopy 1974, Eighth International Congress on Electron Microscopy: Vol. 1* (pp. 1–5). Canberra: Australian Academy of Science.

Ruska, E. (1979). Die frühe Entwicklung der Elektronenlinsen und der Elektronenmikroskopie. *Acta Historica Leopoldina, 12*, 1–136. English translation by Mulvey, T. (1980). *The early development of electron lenses and electron microscopy*. Stuttgart: S. Hirzel Verlag.

Ruska, E. (1986). The emergence of the electron microscope (Connection between realization and first patent application, Documents of an invention). *Journal of Ultrastructure and Molecular Structure Research, 95*, 3–28. See also: Ruska, E. (1984). Die Entstehung des Elektronenmikroskops (Zusammenhang zwischen Realisierung und erster Patentanmeldung, Dokumente einer Erfindung). *Archiv der Geschichte der Naturwissenschaften, 11*, 525–551.

Ruska, E. (1987). The development of the electron microscope and of electron microscopy. Nobel lecture, December 8, 1986. *Reviews of Modern Physics, 59*, 627–638.

Ruska, E., & Knoll, M. (1931). Die magnetische Sammelspule für schnelle Elektronenstrahlen. *Zeitschrift für technische Physik, 12*, 389–400, 448.

Scherzer, O. (1936). Über einige Fehler von Elektronenlinsen. *Zeitschrift für Physik, 101*, 593–603.

Sommerfeld, A., & Scherzer, O. (1934). Über das Elektronenmikroskop. *Münchener Medizinische Wochenschrift, 81*, 1859–1860.

Stintzing, H. (1927). Verfahren und Einrichtung zum automatischen Nachweiss, Messung und Zählung von Einzelteilchen beliebiger Art, Form und Grösse. German patent 485155, filed on 13 May 1927, granted on 28 December 1929.

Süsskind, C. (1985). L.L. Marton, 1901–1979. In P. W. Hawkes (Ed.), *Advances in Electronics and Electron Physics: Supplement 16. The beginnings of electron microscopy* (pp. 501–523). Orlando: Academic Press.

Trillat, J.-J. (1962). Some personal reminiscences. In P. P. Ewald (Ed.), *Fifty Years of X-ray Diffraction* (pp. 662–666). Utrecht: Oosthoek's Uitgeversmaatschappij.

Ura, K. (1996). History of electron microscopes at Osaka University, 1934–1945. In P. W. Hawkes, B. Kazan, & T. Mulvey (Series Eds.), *Advances in Imaging and Electron Physics: Vol. 96. The growth of electron microscopy* (pp. 263–269). San Diego: Academic Press.

van Gorkom, J. (2018). The early electron microscopes: A critical study. *Advances in Imaging and Electron Physics, 205*, 1–137.

van Gorkom, J., & de Haas, B. (1985). Interview with Ernst Ruska in Berlin on 25 June, 27 June and 3 July 1985.

von Borries, B. (1934). Verfahren zur Abbildung von Flächen mittels Korpuskularstrahlen. German patent 692336, priority date of 7 December 1934, granted on 23 May 1940.

von Borries, B. (1935). Achromatische Linse für die Abbildung met Elektronenstrahlen. German patent 721417, priority date of 11 May 1935, granted on 30 April 1942.

von Borries, B., & Ruska, E. (1932a). Das kurze Raumladungsfeld einer Hilfsentladung als Sammellinse für Kathodenstrahlen. *Zeitschrift für Physik, 76*, 649–654.

von Borries, B., & Ruska, E. (1932b). Magnetische Sammellinse kurzer Feldlänge. German patent 680284, priority date of 17 March 1932, published on 3 August 1939.

von Borries, B., & Ruska, E. (1932c). Anordnung zur Beobachtung und Kontrolle der im Strahlengang eines Elektronenmikroskops mit zwei oder mehr elektronenoptischen Vergrösserungsstufen auftretenden elektronenoptischen Bilder. German patent 679857, priority date of 17 March 1932, published on 15 August 1939.

von Borries, B., & Ruska, E. (1933a). Die Abbildung durchstrahlter Folien im Elektronenmikroskop. *Zeitschrift für Physik, 83*, 187–193.

von Borries, B., & Ruska, E. (1933b). Anordnung zur Erzeugung eines parallelen oder kegelförmigen Kathodenstrahlen großer Dichte. German patent 692335, filed on 1 December 1933, granted on 23 May 1940.

von Borries, B., & Ruska, E. (1935). Das Elektronenmikroskop und seine Anwendungen. *Zeitschrift des Vereines Deutscher Ingenieure (a.k.a. VDI-Zeitschrift), 79*, 519–524.

von Borries, B., & Ruska, E. (1936). Angewandte Elektronenoptik. *Zeitschrift des Vereines Deutscher Ingenieure (a.k.a. VDI-Zeitschrift), 80*, 989–994, 1075–1083.

von Borries, H. (1991). Bodo von Borries. *Advances in Electronics and Electron Physics, 81*, 127–176.

von Laue, M. (1946). Arnold Berliner. *Naturwissenschaften, 33*, 257–258.

Watson, J. H. L. (2013). *The electron microscope, a personal recollection.* http://www. physics.utoronto.ca/physics-at-uoft/history/the-electron-microscope/the-electron-microscope-a-personal-recollection/. (Accessed 27 June 2013).

Yada, K. (1996). History of electron microscopes at Tohoku University. In P. W. Hawkes, B. Kazan, & T. Mulvey (Series Eds.), *Advances in Imaging and Electron Optics: Vol. 96. The growth of electron microscopy* (pp. 245–249). San Diego: Academic Press.

Zworykin, V. K. (1933). On electron optics. *Journal of the Franklin Institute, 215*, 535–555.

Zworykin, V. K. (1936). Elektronenoptische Systeme und ihre Anwendung. *Zeitschrift für technische Physik, 17*, 170–183.

CHAPTER THREE

Three Dimensional Computer Modeling of Electron Optical Systems*

John A. Rouse

Applied Optics Section, Blackett Laboratory, Imperial College of Science, Technology, and Medicine, London, United Kingdom
e-mail address: john@mebs.co.uk

Contents

---

* Reprinted from Advances in Optical and Electron Microscopy, vol. 13 (1994) 1–121.

Advances in Imaging and Electron Physics, Volume 208
ISSN 1076-5670
https://doi.org/10.1016/bs.aiep.2018.09.001

1. INTRODUCTION

1.1 Background

The business of building charged particle optical columns is a highlytechnological and costly one. The manufacture of each component must be carried out accurately and the final column must be aligned precisely, evacuated, and driven by complicated electronic circuits, thus making the building of prototypes expensive and time consuming. It has long been recognized that computer modeling of the components of these machines can dramatically cut the cost of each iteration in the design phase, obviating the need to build so many prototype parts. In addition, the computer models often give insight into the physics of the processes occurring which may, in turn, indicate the features of the design which limit the performance of the components (e.g., an area where a polepiece is magnetically saturated, or a region where the electric field between two electrodes is greater than the breakdown value, or where electrons are missing their intended target). This is especially relevant because the detection of these faults is difficult when the components are installed in the column. By being aware of these shortcomings in the design, the designer is able to achieve a required performance specification in fewer iterations than by trial and error methods.

Two qualifications must be added to this: firstly, the computer simulations must be accurate and reliable, so that the designers do not feel obliged to build each modified design to make sure the program is still well behaved; secondly, the cost and time invested in the computer simulations must be much less than that required to build the column by repeated prototype manufacture.

Fifteen years ago Munro (1975) introduced a set of computer programs based on the finite element method (FEM) for analysis of rotationally symmetric electron lenses. These programs represented a fundamental step forward in the field of computer aided design (CAD) of electron optical components because, for the first time, they handled lenses with complicated shapes, dielectric materials, ferromagnetic polepieces including saturation and coil windings in a unified way. In some cases, the solution of several thousand non-linear simultaneous equations was required which, at the time the programs were first developed, took several minutes to run on a large mainframe computer.

Since that time, the rapid pace of computer hardware development has transformed the situation enormously. The latest versions of Munro's 2D FEM programs can now be run in a matter of seconds on a modern, state-of-the-art desk top personal computer (PC) (Munro, 1990), and, indeed, have been installed on such machines at many sites throughout the electron optics industry. This has enabled electron optical designers to use these and similar CAD tools (Lencova & Wisselink, 1990) in an interactive way in their own offices or laboratories, utilizing any of the convenient peripherals these PC machines offer (e.g., digitizers, plotters, and floppy disk drives).

Up till now, most computer aided design programs in electron optics compute the fields in two dimensions (2D). This is possible if one exploits the rotational symmetry of the lenses and expresses the deflection fields in Fourier Harmonic components to reduce the dimensionality of the problem. However, many important applications in electron optics call for a fully three dimensional (3D) field analysis and electron ray-trace. Although some specialized 3D electron optics programs have been developed (Franzen, 1984; MacGregor, 1983; Munro & Chu, 1982; Smith & Munro, 1987), they have mostly been restricted to handling special cases. The aim of the present work was to develop a suite of general purpose 3D electron optical design programs which can be run in a PC environment.

1.2 Basic Aims of a 3D Electron Optical Design Package

To analyze a problem in electron optics, one essentially needs to know how a set of electrons at various positions in one space – the object space – map to a corresponding set of positions in another space – the image space. The relationships between these sets of co-ordinates is primarily determined by the electric and magnetic forces exerted on the electrons as they pass from object to image. Thus, the major task in most computer modeling of such systems is to compute the electric and magnetic fields set up by the electrodes and polepieces in the column. Once this has been achieved, the electron optical properties can be computed either by directly tracing the path of the electrons through the fields or by using the axial and near axial fields in conjunction with an aberration theory to map many electrons from object to image, assuming they do not travel very far from the chosen axis. For many fully 3D problems, a direct electron ray-trace is essential for two reasons: often the assumptions made in the derivation of current aberration theories are violated (for example, third order aberration theories are valid only for small angle deviations from the axis); and in some cases, no general optical axis can be conveniently defined (for example in modeling the collection of secondary electrons from a sample in a scanning electron microscope). Therefore, the next computational task in a 3D analysis package is a direct electron ray-trace.

1.3 Methods of Computing 3D Electromagnetic Field Distributions

There are several methods for computing the fields in electromagnetic systems that are viable but have strong and weak points. There is no single formulation which is the best for every 3D analysis.

The charge density method (CDM) or boundary element method (BEM) (Costabel, 1987; Desbruslais & Munro, 1987) is mathematically elegant. The 3D solid electrodes are replaced by their 2D surfaces and covered with the necessary surface charge density to give the required electrode voltages. Once this charge distribution has been computed, the potentials and fields in all space can be found by a 2D surface integral of the charges. It therefore reduces the 3D problem to one of computing a set of 2D integrals. This method is ideal for general 3D electrostatic systems with large scale factor changes or open boundaries, since the method does not discretize the 3D space over which the solution is required; instead, it discretizes the surfaces over which the charges are placed. The method works best for objects with smooth surfaces (e.g., spheres), but can have problems with surfaces which have many edges, corners or points. In these cases the charge density can be concentrated in critical regions and a lot of the mathematical elegance and simplicity of the method can be lost trying to integrate round these singularities. Furthermore, even when the surface charges have been found, a 2D integral over all the electrode surfaces is required to obtain the field at any point. This can make the electron ray-traces prohibitively slow in some cases, since the electric field is required at every step on the ray and many trajectories are often required.

The finite element method (FEM) (Morton, 1987; Silvester & Ferrari, 1990) is well established in 2D and 3D. It is a very powerful method which can compute fields in electrostatic systems of electrodes and dielectrics, and magnetic systems of coil windings and polepieces, including the effects of saturation. The method discretizes the whole volume of space over which the solution is required into a mesh of elements. In 3D the mesh must be defined so that the surfaces of the electrodes, polepieces, and coil windings lie along the planes joining the mesh nodes. In practice, this requires the construction of a 3D mesh to fit the objects; this can often be quite tricky, requiring several iterations of configuring the mesh, to obtain a suitable grid of nodes. Furthermore, when the potentials have been computed, it is difficult to obtain an accurate estimate of the field at any point, because the potentials are computed at points on an irregular mesh and, therefore, cannot be differenced in a simple way. For fast ray-traces it is often sufficient to assume that the field in each element is constant and equal to some average of the potentials surrounding the element. More complicated schemes have been proposed for computing the fields to a greater accuracy in a 2D FEM mesh (cf. Khursheed & Dinnis, 1989), but these are more time consuming. Recently, the use of second order elements has been reported by Zhu and Munro (1989), where the field is assumed to be quadratic in each element. This provides an order of magnitude improvement in accuracy of the potentials and fields and has been successfully implemented in 2D. The second order elements also allow the curvature of the surfaces to be modeled more accurately. However, even in 2D, these second order FEM programs are much more time and memory consuming than their first order counterparts. Accurate direct ray-tracing can be especially time-consuming. The technique proposed by Zhu and Munro (1989) to obtain the electric fields, involves a coordinate transformation onto a regular square mesh and requires frequent use of many mapping coefficients. In 3D, the mapping of the second order FE mesh onto a regular grid and the interpolation of the potentials would be an order of magnitude more memory and time consuming.

The finite difference method (FDM) (Weber, 1967; Kasper, 1982) is a very old technique and was probably the first numerical method used to compute electrostatic fields in electron optics. It also discretizes the volume of space over which the solution is required into a grid of mesh points but, unlike FEM, the grid is a regular Cartesian (or cylindrical) one. The advantage of this, from the user's viewpoint, is that setting up the mesh data is much simpler, since the specification of the mesh is essentially decoupled from the specification of the electrodes and polepieces. However, from the programmer's viewpoint more work is involved, because the objects (electrodes and polepieces) may intersect the mesh at any position (not only at the nodes), and at sites where an object's surface has cut one of the arms linking two nodes, modifications must be made to the finite difference equations used. When the potentials have been computed, the fields can be obtained quickly and accurately by differencing the potentials on the regular grid. This makes the ray-tracing much quicker than with the FEM or CDM for the same accuracy.

It was felt that, on account of the advantages noted above, the finite difference method still has a lot to offer, especially in the 3D aspects of electron optics. To make a useful design tool the programs should be capable of analyzing electrostatic and magnetic systems including electrodes, dielectrics, surface charges, polepieces, and coil windings of quite general 3D shapes in a unified and integrated way. To do this, special equations must be formulated to handle all these cases in 3D. Furthermore, the programs must be arranged to be as fast and memory efficient as possible so as to be able to run on the small computers which can now be a part of every electron optics designer's office.

2. FORMULATING THE FINITE DIFFERENCE EQUATIONS

2.1 Introduction

The finite difference method (FDM) is an old technique, first systematically described by Sheppard (1899) at the turn of the century, but whose roots date back to the very origin of differential calculus. Before the advent of modern computing facilities, the FDM was used to provide equations which were solved by hand using the relaxation methods of Southwell (1940). Relaxation was first proposed in 1935 for calculating stresses in engineering frameworks and obviated the need to solve large numbers of simultaneous equations directly. The method was later used by Southwell (1946) and Allen (1954) in the solution of Laplace's equation formulated using the FDM. At this early stage, the relaxation was a task requiring specialized skills, since the order in which the various equations were solved depended on the solver looking at where the significant changes were occurring and concentrating on these areas. As electronic computers were developed in the 1950s, the successive over-relaxation (SOR) method was developed by Frankel (1950) and Young (1954) to solve the finite difference equations. The combination of FDM and SOR was used in the early days of numerical analysis in electron optics for studying electrostatic lenses and magnetic lenses in the polepiece region using a scalar potential (e.g., Kamminga, Verster, & Franken, 1968). However, since the introduction of the finite element method (FEM) to electron optics by Munro (1971) in the early 1970s, the use of FDM has dwindled – especially for the analysis of magnetic lenses or dielectric materials. The FDM still has several advantages over FEM, especially in three-dimensional analyses (as discussed in Section 1), but for the method to be able to compete with the FEM, it must be able to handle full 3D magnetic systems including the coil windings and also 3D dielectric materials with surface charges – neither of which has been possible before. In this section, the theory of the FDM and SOR methods are reviewed and a new theory is presented which enables the FDM to be used to solve complicated 3D systems of electrodes, dielectrics, and unsaturated magnetic materials including the coil windings in a unified way.

2.2 Basic Formulation of the Finite Difference Method

This section reviews the basic concepts of the finite difference method and the over-relaxation technique used to solve the large sets of equations generated by such methods.

2.2.1 Discretized Form of Laplace's Equation

To solve a partial differential equation such as Laplace's equation,

$$\nabla^2 \Phi = 0 \tag{1}$$

numerically using the finite difference method (FDM), the volume of space throughout which the solution is required must first be discretized into a grid of nodes. At each of these nodes the function, Φ, is defined and a form of Eq. (1) must be found which gives a relation between the value of Φ at any node and the values of Φ at its nearest neighboring nodes. Such an equation is known as a difference equation. Consider Eq. (1) in its three dimensional (3D) Cartesian form:

$$\frac{\partial^2 \Phi}{\partial x^2} + \frac{\partial^2 \Phi}{\partial y^2} + \frac{\partial^2 \Phi}{\partial z^2} = 0 \tag{2}$$

The region in which Laplace's equation is to be solved is covered by a Cartesian grid of points. Let Φ_0 (Fig. 1) be the potential at a general node on the grid, and $\Phi_1 - \Phi_6$ be the potentials at the six adjacent nodes in x, y, and z directions, as shown, at distances $h_1 - h_6$ from the center node. To obtain an approximation for the value of $\partial^2 \Phi / \partial x^2$, the potentials Φ_1 and Φ_2 are expressed by a Taylor series expansion in the x-direction up to second order terms:

$$\Phi_1 = \Phi_0 + \left. \frac{\partial \Phi}{\partial x} \right|_0 h_1 + \frac{1}{2} \left. \frac{\partial^2 \Phi}{\partial x^2} \right|_0 h_1^2 + \ldots \tag{3}$$

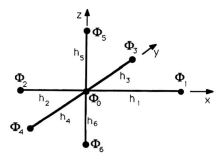

Figure 1 A single node with potential Φ_0 in the 3D finite difference mesh and its six nearest neighbors at distances $h_1 \dots h_6$, with potentials $\Phi_1 \dots \Phi_6$.

Table 1 Formulæ for the coefficients of the standard seven-point 3D FD equation

$\alpha_1 = 2/[h_1(h_1 + h_2)]$	$\alpha_2 = 2/[h_2(h_1 + h_2)]$
$\alpha_3 = 2/[h_3(h_3 + h_4)]$	$\alpha_4 = 2/[h_4(h_3 + h_4)]$
$\alpha_5 = 2/[h_5(h_5 + h_6)]$	$\alpha_6 = 2/[h_6(h_5 + h_6)]$
$\alpha_0 = \alpha_1 + \alpha_2 + \alpha_3 + \alpha_4 + \alpha_5 + \alpha_6$	

$$\Phi_2 = \Phi_0 - \left.\frac{\partial \Phi}{\partial x}\right|_0 h_2 + \frac{1}{2}\left.\frac{\partial^2 \Phi}{\partial x^2}\right|_0 h_2^2 - \dots \tag{4}$$

We can then solve for $\partial^2\Phi/\partial x^2|_0$ in terms of Φ_0, Φ_1, and Φ_2 by eliminating the first derivative term, $\partial\Phi/\partial x|_0$.

By applying the same principle in the y and z directions in Fig. 1, expressions for the second partial derivatives of Φ with respect to y and z at node 0 can be derived and substituted into Laplace's equation (2). The final difference equation can be written in the general form:

$$\boxed{\alpha_1\Phi_1 + \alpha_2\Phi_2 + \alpha_3\Phi_3 + \alpha_4\Phi_4 + \alpha_5\Phi_5 + \alpha_6\Phi_6 = \alpha_0\Phi_0} \tag{5}$$

where the coefficients $\alpha_0 \dots \alpha_6$ are given in Table 1.

Eq. (5) is the difference form of Laplace's equation and relates the value of Φ at any node to the values of Φ at its nearest neighboring nodes, as required.

The procedure for solving an electrostatic potential distribution around a set of electrodes is to set up a 3D mesh of nodes at each of which a potential, Φ, is defined. Nodes which are situated inside the electrode regions are fixed at the electrode potentials and form the boundary conditions for

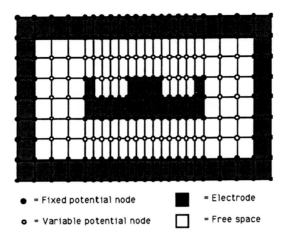

● = Fixed potential node   ■ = Electrode

○ = Variable potential node   □ = Free space

Figure 2 2D rectangular electrode geometry in a finite difference mesh with fixed and variable potential nodes indicated.

the problem. The potentials at the remaining nodes are unknown and are the variables in the problem.

2.2.2 Treatment of Curved Electrodes

Fig. 2 gives a 2D example of the discretization of an electrode geometry and the assignment of nodes. If the definitions of the h's in Eq. (5) are used (i.e. that they are the distances between consecutive nodes), then only electrodes whose shapes conform to the rectangular grid of nodes can be solved for, using the formulæ derived thus far. In the 3D case, this means that the electrodes must consist entirely of flat surfaces, parallel to the three principal planes of the Cartesian coordinate system. Furthermore, the 3D Cartesian finite difference grid must be chosen so that all the electrode surfaces pass exactly through the grid points.

In most 3D analyses, the structure of the electrodes cannot be broken down into a set of elements with flat sides parallel to, and coincident with, the Cartesian grid. In these cases, the electrode surfaces may cut the mesh lines of the finite difference grid and Eq. (5) can no longer be applied at nodes where this occurs. To allow for this possibility, a minor modification in the interpretation of the h's of Eq. (5) must be made.

Consider an electrode at potential Φ_e, whose surface cuts one of the arms of the mesh (Fig. 3). Instead of h_i representing the distance from node 0 to node i, it is now interpreted as the distance from 0 to the

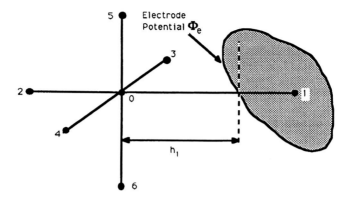

Figure 3 Special case of a node adjacent to an electrode surface with modified arm length, h_1.

electrode cut along the affected arm, and Φ_i is replaced by the electrode potential Φ_e. Using these definitions, Eq. (5) can still be used at each variable point, since the modifications ensure that the FD equation is still applied in an isotropic region in which Laplace's equation is valid.

2.2.3 Solving the Equations

Once the grid has been set up, the difference equation (5) is then applied at each of the nodes outside the electrodes. If there are N such nodes, then the problem is essentially one of solving N linear equations in N unknowns. For 3D problems, especially, N is large (typically several tens of thousands). Several established techniques exist for solving a large matrix of simultaneous equations.

A direct method of solution is Gaussian elimination, which has been used to great effect in 2D finite element programs (e.g., Munro, 1971). In the finite element method (FEM), it is possible to arrange the mesh so that the matrix to be solved is symmetric and banded (i.e., the matrix has zeros everywhere, except on the diagonal and a few lines parallel to the diagonal). An efficient solving algorithm can be used in this 2D case. In the FDM, although the matrix may be sparse (i.e., lots of zero coefficients in it), the non–zero elements are not regularly arranged as they are in the FEM. Furthermore, the bandwidth of the non–zero elements is much greater in the 3D than the 2D case, which would require much greater computational time, on account of the filling of the band with non–zero elements during the Gaussian elimination. To solve a full 3D FDM matrix equation by

Gaussian elimination would therefore prove prohibitively slow and memory consuming on a personal computer.

Another solution method is the Incomplete Choleski Conjugate Gradient (ICCG) method (Meijerink & Van der Vorst, 1977), which was introduced to electron optics by Lencova and Lenc (1984) to solve large 2D FEM matrix equations. It is a semi–iterative technique and, for large matrices, requires less storage and (usually) less computation time than the Gaussian elimination method. The ICCG method was originally formulated to solve symmetric matrices, whereas the FDM matrix is non–symmetric. (Later ICCG schemes Jacobs, 1983 have been reported for asymmetric FDM matrices.) However, in 3D even the ICCG method requires about 15 variables per mesh point.

The method of successive over-relaxation (SOR) was chosen to solve the 3D FDM matrix equations. It is a relatively simple technique, requiring only one variable per mesh point (except at nodes where an electrode surface cuts one of the nodal arms – e.g., as shown for arm 1 in Fig. 3).

2.3 A New Formulation for Dielectric and Magnetic Materials

So far only electrostatic systems containing 3D electrodes have been considered. However, many electron optical systems contain insulating parts or polepieces consisting of magnetically permeable iron. Although, in most cases, non–conducting elements of the column are kept away from the electron beam, in some cases this is not possible (for example when inspecting insulating specimens in an electron microscope). Furthermore, the handling of magnetic materials, including the coil windings, is crucial to the complete modeling of electron beam columns. To study such problems, it has been necessary to develop a new extension to the basic finite difference method which takes account of dielectric and permeable materials.

For 2D problems, a formulation using an analogy with a rubber membrane to model the refraction of magnetic fields at the surface of iron has been described by Southwell (1946). In this method, the grid lines of the finite difference mesh are thought of as elastic strings which form a 2D "net" covering the region of interest. The relaxation equations are formed by considering the tensions in each string connected to any two nodes, when the net is deflected by a piece of permeable material placed on it. For three dimensions, it was found that an extension of this "string-net" analogy was inappropriate, because if the mesh is already three-dimensional, one would have to imagine that the presence of the iron would deflect the

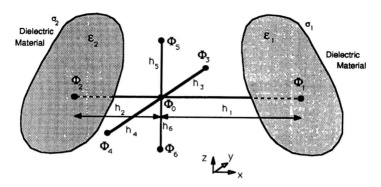

Figure 4 Finite difference star with dielectric material interfaces on each arm of the x-axis.

"net" in some fourth dimension. Although this may be mathematically feasible, it was felt that much of the simplicity of the physical analogy with a net of tensioned strings would be lost. Therefore, a new formulation using finite differences is described in the following sections, which models 3D dielectric materials, which may have any applied surface charge distributions. This is then extended to deal with linear magnetic systems (i.e., non–saturated polepieces), including the presence of the coil windings.

2.3.1 New Theory of FD Equations for Single Dielectric Cuts

Initially we assume that the dielectric surface cuts any one arm of the 3D mesh, at most, only once. Consider a node 0 at potential Φ_0 (Fig. 4). Suppose the mesh line in the positive x-direction in the vicinity of 0 (arm 1) has been cut by the surface of a dielectric material of relative permittivity ϵ_{r1}, which has a surface charge density of σ_1 C/m². Also, suppose the mesh line in the negative x-direction in the vicinity of 0 (arm 2) has been cut by the surface of a dielectric material of relative permittivity ϵ_{r2}, which has a surface charge density of σ_2 C/m².

Let these intersection points with the arms be at distances a_1 and a_2 from 0, and have normals of n_1 and n_2 respectively (Fig. 5).

We require a difference equation relating the potential at node 0 to the potentials of its nearest neighboring nodes in the form of Eq. (5). This equation cannot be used directly if one or more of the arms passes through a dielectric interface, since Laplace's equation does not hold at the interface and the potential there is unknown. In this case we need to model the field discontinuity at the dielectric interface. As in the basic formulation of the

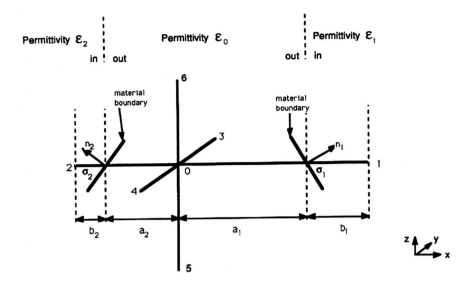

Figure 5 Finite difference star with dielectric cuts on arms 1 and 2.

FDM (see Section 2.2), we will again consider the Taylor series expansion of the potential up to second order terms, so as to obtain the same truncation errors as before. In the x-direction, therefore, we expand the potential, Φ, about 0 in a Taylor series up to terms in x^2, and differentiate twice with respect to x, retaining terms up to second order in x:

$$\Phi(x) = \Phi_0 + \Phi_x x + \frac{1}{2}\Phi_{xx}x^2$$
$$\Phi'(x) = \Phi_x + \Phi_{xx}x$$
$$\Phi''(x) = \Phi_{xx}$$
$$(-a_2 \leq x \leq a_1)$$

where:
- $\Phi(x)$ is the potential at x, $-\Phi'(x)$ is the x-field at x, $-\Phi''(x)$ is the gradient of the x-field at x;
- Φ_0 is the potential at 0, $-\Phi_x$ is the x-field at 0, $-\Phi_{xx}$ is the gradient of the x-field at 0.

(N.B. The "$-$" signs arise because the electric field, **E**, is given by: $\mathbf{E} = -\nabla\Phi$.)

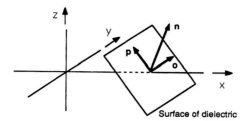

Figure 6 Diagram of the vectors normal and tangential to the surface of the dielectric material at the intersection point with the FD mesh.

Therefore, just **outside** the material we have at the right and left cut respectively:

$$\Phi(a_1 - \delta) = \Phi_0 + \Phi_x a_1 + \frac{1}{2}\Phi_{xx}a_1{}^2$$
$$\Phi'(a_1 - \delta) = \Phi_x + \Phi_{xx}a_1$$
$$\Phi''(a_1 - \delta) = \Phi_{xx} \tag{6}$$

and

$$\Phi(-a_2 + \delta) = \Phi_0 - \Phi_x a_2 + \frac{1}{2}\Phi_{xx}a_2{}^2$$
$$\Phi'(-a_2 + \delta) = \Phi_x - \Phi_{xx}a_2$$
$$\Phi''(-a_2 + \delta) = \Phi_{xx} \tag{7}$$

(where δ is a small distance, i.e., $\delta << a_1$, $\delta << a_2$).

We now want to write similar equations for Φ, Φ', and Φ'' just **inside** the material at the cuts. We can immediately write down the value of Φ since the potential is the same on either side of each cut:

$$\Phi(a_1 + \delta) = \Phi_0 + \Phi_x a_1 + \frac{1}{2}\Phi_{xx}a_1{}^2 \tag{8}$$

and

$$\Phi(-a_2 - \delta) = \Phi_0 - \Phi_x a_2 + \frac{1}{2}\Phi_{xx}a_2{}^2 \tag{9}$$

To obtain an expression for Φ' we resolve the field in directions normal (**n**) and tangential (**o** and **p**) to the surface at the cut (see Fig. 6):

$$\Phi'(a_1 + \delta) = \Phi_{n1}n_{x1} + \Phi_{o1}o_{x1} + \Phi_{p1}p_{x1} \tag{10}$$

and

$$\Phi'(-a_2 - \delta) = \Phi_{n2}n_{x2} + \Phi_{o2}o_{x2} + \Phi_{p2}p_{x2} \tag{11}$$

where (see Fig. 6):

- $-\Phi_{n1}$ is the normal component and $-\Phi_{o1}$ and $-\Phi_{p1}$ are the tangential components of the field at the cut on arm 1.
- n_1 is a unit vector normal to the surface of the cut on arm 1, in the direction away from node 0 (see Fig. 5), and o_1 and p_1 are unit vectors tangential to the surface at the cut on arm 1.
- $-\Phi_{n2}$ is the normal component and $-\Phi_{o2}$ and $-\Phi_{p2}$ are the tangential components of the field at the cut on arm 2.
- n_2 is a unit vector normal to the surface of the cut on arm 2, in the direction away from node 0 (see Fig. 5), and o_2 and p_2 are unit vectors tangential to the surface at the cut on arm 2.

As the surface is crossed, the outward normal component of the electric displacement, $\mathbf{D}$, jumps by an amount σ, where σ is the surface charge density. For example, on arm 1,

$$D_{n1}^{in} = D_{n1}^{out} + \sigma_1$$

i.e.,

$$\epsilon_0 \epsilon_{r1} E_{n1}^{in} = \epsilon_0 \epsilon_{r0} E_{n1}^{out} + \sigma_1 \tag{12}$$

where ϵ_0 is the electric constant (8.855×10^{-12} F/m), ϵ_{r0} and ϵ_{r1} are the relative permittivities at nodes 0 and 1 respectively (see Fig. 5), and E_{n1}^{in} and E_{n1}^{out} are the normal components of the electrostatic field, on arm 1, just inside and just outside of the dielectric interface (see Fig. 5).

Since $\mathbf{E} = -\nabla\Phi$ (i.e., $E_n = -\Phi_n$), Eq. (12) can be expressed as

$$\Phi_{n1}^{in} = \frac{\epsilon_{r0}}{\epsilon_{r1}} \Phi_{n1}^{out} - \frac{\sigma_1}{\epsilon_0 \epsilon_{r1}} \tag{13}$$

and similarly for arm 2 (see Fig. 5), we obtain

$$\Phi_{n2}^{in} = \frac{\epsilon_{r0}}{\epsilon_{r2}} \Phi_{n2}^{out} - \frac{\sigma_2}{\epsilon_0 \epsilon_{r2}} \tag{14}$$

[Note that ϵ_0 denotes the electric constant (8.855×10^{-12} F/m), and ϵ_{r0} means the relative permittivity at node 0.]

Also, since the tangential components of Φ are continuous at the cuts (i.e., $\Phi_{o1}^{in} = \Phi_{o1}^{out}$, $\Phi_{p1}^{in} = \Phi_{p1}^{out}$), we have from Eqs. (10) and (11) respectively,

$$\Phi'(a_1 + \delta) = \left(\frac{\epsilon_{r0}}{\epsilon_{r1}} \Phi_{n1}^{out} - \frac{\sigma_1}{\epsilon_0 \epsilon_{r1}} \right) n_{x1} + \Phi_{o1}^{out} o_{x1} + \Phi_{p1}^{out} p_{x1} \qquad (15)$$

and

$$\Phi'(-a_2 - \delta) = \left(\frac{\epsilon_{r0}}{\epsilon_{r2}} \Phi_{n2}^{out} - \frac{\sigma_2}{\epsilon_0 \epsilon_{r2}} \right) n_{x2} + \Phi_{o2}^{out} o_{x2} + \Phi_{p2}^{out} p_{x2} \qquad (16)$$

We now **assume** that Φ_y and Φ_z are independent of x in the region $-a_2 < x < a_1$ (i.e., the values of Φ_y and Φ_z just inside the cuts are the same as at the center node, 0). This is the best approximation we can make because, with a 7-point equation, we have no method of estimating the mixed partial derivative terms (e.g., $\partial^2 \Phi / \partial x \partial y$).

Then,

$$\Phi_{n1}^{out} = \Phi'(a_1 - \delta) n_{x1} + \Phi_y n_{y1} + \Phi_z n_{z1}$$
$$\Phi_{o1}^{out} = \Phi'(a_1 - \delta) o_{x1} + \Phi_y o_{y1} + \Phi_z o_{z1}$$
$$\Phi_{p1}^{out} = \Phi'(a_1 - \delta) p_{x1} + \Phi_y p_{y1} + \Phi_z p_{z1}$$

and

$$\Phi_{n2}^{out} = \Phi'(-a_2 + \delta) n_{x2} + \Phi_y n_{y2} + \Phi_z n_{z2}$$
$$\Phi_{o2}^{out} = \Phi'(-a_2 + \delta) o_{x2} + \Phi_y o_{y2} + \Phi_z o_{z2}$$
$$\Phi_{p2}^{out} = \Phi'(-a_2 + \delta) p_{x2} + \Phi_y p_{y2} + \Phi_z p_{z2}$$

Substituting these into Eqs. (15) and (16) respectively and rearranging, we can write down an expression for the field just inside the dielectrics:

$$\Phi'(a_1 + \delta) = \left[\frac{\epsilon_{r0}}{\epsilon_{r1}} n_{x1}^2 + o_{x1}^2 + p_{x1}^2 \right] \Phi'(a_1 - \delta)$$
$$+ \left[\frac{\epsilon_{r0}}{\epsilon_{r1}} n_{x1} n_{y1} + o_{x1} o_{y1} + p_{x1} p_{y1} \right] \Phi_y$$
$$+ \left[\frac{\epsilon_{r0}}{\epsilon_{r1}} n_{x1} n_{z1} + o_{x1} o_{z1} + p_{x1} p_{z1} \right] \Phi_z - \frac{\sigma_1 n_{x1}}{\epsilon_0 \epsilon_{r1}} \qquad (17)$$

and

$$\Phi'(-a_2 - \delta) = \left[\frac{\epsilon_{r0}}{\epsilon_{r2}} n_{x2}^2 + o_{x2}^2 + p_{x2}^2 \right] \Phi'(-a_2 + \delta)$$

$$+ \left[\frac{\epsilon_{r0}}{\epsilon_{r2}} n_{x2} n_{y2} + o_{x2} o_{y2} + p_{x2} p_{y2} \right] \Phi_y$$

$$+ \left[\frac{\epsilon_{r0}}{\epsilon_{r2}} n_{x2} n_{z2} + o_{x2} o_{z2} + p_{x2} p_{z2} \right] \Phi_z - \frac{\sigma_2 n_{x2}}{\epsilon_0 \epsilon_{r2}} \qquad (18)$$

Now, since n, o, and p are orthogonal unit vectors, it can be shown that $n_{x1}^2 + o_{x1}^2 + p_{x1}^2 = 1$ (and similarly for y and z), and that $n_{x1} n_{y1} + o_{x1} o_{y1} + p_{x1} p_{y1} = 0$ (and similarly for other unlike combinations of x, y, and z). These relations enable the square bracketed terms in Eqs. (17) and (18) to be simplified, and the equations can be written in the following form:

$$\Phi'(a_1 + \delta) = [\gamma_1 n_{x1}^2 + 1] \Phi'(a_1 - \delta) + \gamma_1 n_{x1} n_{y1} \Phi_y + \gamma_1 n_{x1} n_{z1} \Phi_z - \frac{\sigma_1 n_{x1}}{\epsilon_0 \epsilon_{r1}} \quad (19)$$

and

$$\Phi'(-a_2 - \delta) = [\gamma_2 n_{x2}^2 + 1] \Phi'(-a_2 + \delta) + \gamma_2 n_{x2} n_{y2} \Phi_y + \gamma_2 n_{x2} n_{z2} \Phi_z - \frac{\sigma_2 n_{x2}}{\epsilon_0 \epsilon_{r2}}$$

$$(20)$$

where $\gamma_i = \left(\frac{\epsilon_{r0}}{\epsilon_{ri}} - 1 \right)$.

For the second derivative just inside the dielectrics we **assume**:

$$\Phi''(a_1 + \delta) = \Phi_{xx} \qquad (21)$$

and

$$\Phi''(-a_2 - \delta) = \Phi_{xx} \qquad (22)$$

This approximation is valid for cuts which are normal to or parallel to the axis, but does not hold exactly for cuts at an oblique angle to the axis. Again, this approximation is forced on us because the seven point star (Fig. 1) does not allow a better estimation of the derivatives to be made. We shall see in Section 4, however, that the potentials can be computed to a reasonable accuracy in spite of this approximation.

We now have expressions for Φ, Φ', and Φ'' just outside the cuts, and we can express the potentials on the ends of the arms in terms of these. We have, by Taylor expansion of Φ at the cuts,

$$\Phi_1 = \Phi(a_1 + \delta) + b_1 \Phi'(a_1 + \delta) + \frac{1}{2} b_1^2 \Phi''(a_1 + \delta)$$

and

$$\Phi_2 = \Phi(-a_2 - \delta) + b_2 \Phi'(-a_2 - \delta) + \frac{1}{2} b_2^2 \Phi''(-a_2 - \delta)$$

Substituting for Φ, Φ', and Φ'' from Eqs. (6), (7), (8), (9), (19), (20), (21), and (22) gives:

$$\Phi_1 = \Phi_0 + \Phi_x a_1 + \frac{1}{2}\Phi_{xx}a_1{}^2$$
$$+ b_1\left\{[\gamma_1 n_{x1}^2 + 1](\Phi_x + \Phi_{xx}a_1) + \gamma_1 n_{x1} n_{y1}\Phi_y + \gamma_1 n_{x1} n_{z1}\Phi_z - \frac{\sigma_1 n_{x1}}{\epsilon_0 \epsilon_{r1}}\right\}$$
$$+ \frac{1}{2}b_1^2\Phi_{xx}$$

and

$$\Phi_2 = \Phi_0 - \Phi_x a_2 + \frac{1}{2}\Phi_{xx}a_2{}^2$$
$$- b_2\left\{[\gamma_2 n_{x2}^2 + 1](\Phi_x - \Phi_{xx}a_2) + \gamma_2 n_{x2} n_{y2}\Phi_y + \gamma_2 n_{x2} n_{z2}\Phi_z - \frac{\sigma_2 n_{x2}}{\epsilon_0 \epsilon_{r2}}\right\}$$
$$+ \frac{1}{2}b_2^2\Phi_{xx}$$

Grouping the terms and setting $h_1 = a_1 + b_1$, $h_2 = a_2 + b_2$ (where h_1 and h_2 are the arm lengths to mesh-points 1 and 2 respectively) gives:

$$\Phi_1 = \Phi_0 + (h_1 + b_1\gamma_1 n_{x1}^2)\Phi_x + b_1\gamma_1 n_{x1} n_{y1}\Phi_y + b_1\gamma_1 n_{x1} n_{z1}\Phi_z$$
$$+ \left(\frac{1}{2}h_1^2 + a_1 b_1 \gamma_1 n_{x1}^2\right)\Phi_{xx} - \frac{b_1 \sigma_1 n_{x1}}{\epsilon_0 \epsilon_{r1}} \tag{23}$$

and

$$\Phi_2 = \Phi_0 - (h_2 + b_2\gamma_2 n_{x2}^2)\Phi_x - b_2\gamma_2 n_{x2} n_{y2}\Phi_y - b_2\gamma_2 n_{x2} n_{z2}\Phi_z$$
$$+ \left(\frac{1}{2}h_2^2 + a_2 b_2 \gamma_2 n_{x2}^2\right)\Phi_{xx} - \frac{b_1 \sigma_2 n_{x2}}{\epsilon_0 \epsilon_{r2}} \tag{24}$$

Considering the equations for each of the six arms in a similar way, we can write:

$$\Phi_1 - \Phi_0 = R_{1,1}\,\Phi_x + R_{1,2}\,\Phi_y + R_{1,3}\,\Phi_z + S_1\,\Phi_{xx} - Q_1$$
$$\Phi_2 - \Phi_0 = -R_{2,1}\,\Phi_x - R_{2,2}\,\Phi_y - R_{2,3}\,\Phi_z + S_2\,\Phi_{xx} + Q_2$$
$$\Phi_3 - \Phi_0 = R_{3,1}\,\Phi_x + R_{3,2}\,\Phi_y + R_{3,3}\,\Phi_z + S_3\,\Phi_{yy} - Q_3$$
$$\Phi_4 - \Phi_0 = -R_{4,1}\,\Phi_x - R_{4,2}\,\Phi_y - R_{4,3}\,\Phi_z + S_4\,\Phi_{yy} + Q_4$$
$$\Phi_5 - \Phi_0 = R_{5,1}\,\Phi_x + R_{5,2}\,\Phi_y + R_{5,3}\,\Phi_z + S_5\,\Phi_{zz} - Q_5$$
$$\Phi_6 - \Phi_0 = -R_{6,1}\,\Phi_x - R_{6,2}\,\Phi_y - R_{6,3}\,\Phi_z + S_6\,\Phi_{zz} + Q_6 \tag{25}$$

Table 2 Formulæ for the coefficients R, S, and Q in terms of the mesh and surface parameters

i	$R_{i,1}$	$R_{i,2}$	$R_{i,3}$	S_i	Q_i
1	$h_1 + b_1\gamma_1 n_{x1}^2$	$b_1\gamma_1 n_{x1} n_{y1}$	$b_1\gamma_1 n_{x1} n_{z1}$	$\frac{1}{2}h_1^2 + a_1 b_1\gamma_1 n_{x1}^2$	$\frac{b_1\sigma_1 n_{x1}}{\epsilon_0\epsilon_{r1}}$
2	$h_2 + b_2\gamma_2 n_{x2}^2$	$b_2\gamma_2 n_{x2} n_{y2}$	$b_2\gamma_2 n_{x2} n_{z2}$	$\frac{1}{2}h_2^2 + a_2 b_2\gamma_2 n_{x2}^2$	$\frac{b_2\sigma_2 n_{x2}}{\epsilon_0\epsilon_{r2}}$
3	$b_3\gamma_3 n_{x3} n_{y3}$	$h_3 + b_3\gamma_3 n_{y3}^2$	$b_3\gamma_3 n_{y3} n_{z3}$	$\frac{1}{2}h_3^2 + a_3 b_3\gamma_3 n_{y3}^2$	$\frac{b_3\sigma_3 n_{y3}}{\epsilon_0\epsilon_{r3}}$
4	$b_4\gamma_4 n_{x4} n_{y4}$	$h_4 + b_4\gamma_4 n_{y4}^2$	$b_4\gamma_4 n_{y4} n_{z4}$	$\frac{1}{2}h_4^2 + a_4 b_4\gamma_4 n_{y4}^2$	$\frac{b_4\sigma_4 n_{y4}}{\epsilon_0\epsilon_{r4}}$
5	$b_5\gamma_5 n_{x5} n_{z5}$	$b_5\gamma_5 n_{y5} n_{z5}$	$h_5 + b_5\gamma_5 n_{z5}^2$	$\frac{1}{2}h_5^2 + a_5 b_5\gamma_5 n_{z5}^2$	$\frac{b_5\sigma_5 n_{z5}}{\epsilon_0\epsilon_{r5}}$
6	$b_6\gamma_6 n_{x6} n_{z6}$	$b_6\gamma_6 n_{y6} n_{z6}$	$h_6 + b_6\gamma_6 n_{z6}^2$	$\frac{1}{2}h_6^2 + a_6 b_6\gamma_5 n_{z6}^2$	$\frac{b_6\sigma_6 n_{z6}}{\epsilon_0\epsilon_{r6}}$

where the coefficients R, S, and Q are given in Table 2. The subscripts 1–6 in Table 2 refer respectively to the arms 1–6, as illustrated in Fig. 1.

Eqs. (25) are 6 equations in 6 unknowns, namely: Φ_x, Φ_y, Φ_z, Φ_{xx}, Φ_{yy}, Φ_{zz}. Our aim is now to eliminate these unknowns and to form an equation relating the potential at the center node, Φ_0 (see Fig. 1), to the potentials at its six nearest neighboring nodes, $\Phi_1 \dots \Phi_6$.

Eqs. (25) can be written in matrix form:

$$
\begin{pmatrix} \Phi_1 - \Phi_0 \\ \Phi_2 - \Phi_0 \\ \Phi_3 - \Phi_0 \\ \Phi_4 - \Phi_0 \\ \Phi_5 - \Phi_0 \\ \Phi_6 - \Phi_0 \end{pmatrix} = \begin{pmatrix} R_{1,1} & R_{1,2} & R_{1,3} \\ -R_{2,1} & -R_{2,2} & -R_{2,3} \\ R_{3,1} & R_{3,2} & R_{3,3} \\ -R_{4,1} & -R_{4,2} & -R_{4,3} \\ R_{5,1} & R_{5,2} & R_{5,3} \\ -R_{6,1} & -R_{6,2} & -R_{6,3} \end{pmatrix} \begin{pmatrix} \Phi_x \\ \Phi_y \\ \Phi_z \end{pmatrix}
$$

$$
+ \begin{pmatrix} S_1 & S_2 & S_3 & S_4 & S_5 & S_6 \end{pmatrix} \begin{pmatrix} \Phi_{xx} \\ \Phi_{xx} \\ \Phi_{yy} \\ \Phi_{yy} \\ \Phi_{zz} \\ \Phi_{zz} \end{pmatrix} + \begin{pmatrix} -Q_1 \\ Q_2 \\ -Q_3 \\ Q_4 \\ -Q_5 \\ Q_6 \end{pmatrix} \tag{26}
$$

If we define $\boxed{P_i = 1/S_i \ (i = 1, 2, \dots 6)}$ and multiply through by P_i ($i = 1 \dots 6$) then,

$$
\begin{pmatrix}
P_1(\Phi_1 - \Phi_0) \\
P_2(\Phi_2 - \Phi_0) \\
P_3(\Phi_3 - \Phi_0) \\
P_4(\Phi_4 - \Phi_0) \\
P_5(\Phi_5 - \Phi_0) \\
P_6(\Phi_6 - \Phi_0)
\end{pmatrix}
=
\begin{pmatrix}
P_1 R_{1,1} & P_1 R_{1,2} & P_1 R_{1,3} \\
-P_2 R_{2,1} & -P_2 R_{2,2} & -P_2 R_{2,3} \\
P_3 R_{3,1} & P_3 R_{3,2} & P_3 R_{3,3} \\
-P_4 R_{4,1} & -P_4 R_{4,2} & -P_4 R_{4,3} \\
P_5 R_{5,1} & P_5 R_{5,2} & P_5 R_{5,3} \\
-P_6 R_{6,1} & -P_6 R_{6,2} & -P_6 R_{6,3}
\end{pmatrix}
\begin{pmatrix}
\Phi_x \\
\Phi_y \\
\Phi_z
\end{pmatrix}
$$

$$
+
\begin{pmatrix}
\Phi_{xx} \\
\Phi_{xx} \\
\Phi_{yy} \\
\Phi_{yy} \\
\Phi_{zz} \\
\Phi_{zz}
\end{pmatrix}
+
\begin{pmatrix}
-P_1 Q_1 \\
P_2 Q_2 \\
-P_3 Q_3 \\
P_4 Q_4 \\
-P_5 Q_5 \\
P_6 Q_6
\end{pmatrix}
\tag{27}
$$

We now eliminate the Φ_{xx}, Φ_{yy}, and Φ_{zz} terms in two ways:

1. Subtract row 2 from row 1, row 4 from row 3, and row 6 from row 5 in Eq. (27):

$$
\begin{pmatrix}
P_1(\Phi_1 - \Phi_0) - P_2(\Phi_2 - \Phi_0) \\
P_3(\Phi_3 - \Phi_0) - P_4(\Phi_4 - \Phi_0) \\
P_5(\Phi_5 - \Phi_0) - P_6(\Phi_6 - \Phi_0)
\end{pmatrix}
=
$$

$$
\overbrace{
\begin{pmatrix}
P_1 R_{1,1} + P_2 R_{2,1} & P_1 R_{1,2} + P_2 R_{2,2} & P_1 R_{1,3} + P_2 R_{2,3} \\
P_3 R_{3,1} + P_4 R_{4,1} & P_3 R_{3,2} + P_4 R_{4,2} & P_3 R_{3,3} + P_4 R_{4,3} \\
P_5 R_{5,1} + P_6 R_{6,1} & P_5 R_{5,2} + P_6 R_{6,2} & P_5 R_{5,3} + P_6 R_{6,3}
\end{pmatrix}
}^{\mathbf{C}}
\begin{pmatrix}
\Phi_x \\
\Phi_y \\
\Phi_z
\end{pmatrix}
$$

$$
-
\begin{pmatrix}
P_1 Q_1 + P_2 Q_2 \\
P_3 Q_3 + P_4 Q_4 \\
P_5 Q_5 + P_6 Q_6
\end{pmatrix}
\tag{28}
$$

therefore

$$
\mathbf{C}
\begin{pmatrix}
\Phi_x \\
\Phi_y \\
\Phi_z
\end{pmatrix}
=
\begin{pmatrix}
P_1(\Phi_1 - \Phi_0) - P_2(\Phi_2 - \Phi_0) + P_1 Q_1 + P_2 Q_2 \\
P_3(\Phi_3 - \Phi_0) - P_4(\Phi_4 - \Phi_0) + P_3 Q_3 + P_4 Q_4 \\
P_5(\Phi_5 - \Phi_0) - P_6(\Phi_6 - \Phi_0) + P_5 Q_5 + P_6 Q_6
\end{pmatrix}
$$

and if $\mathbf{D}$ is the inverse of $\mathbf{C}$, i.e. $\mathbf{D} = \mathbf{C}^{-1}$, then

$$
\begin{pmatrix} \Phi_x \\ \Phi_y \\ \Phi_z \end{pmatrix} = \begin{pmatrix} D_{1,1} & D_{1,2} & D_{1,3} \\ D_{2,1} & D_{2,2} & D_{2,3} \\ D_{3,1} & D_{3,2} & D_{3,3} \end{pmatrix}
$$
$$
\times \begin{pmatrix} P_1(\Phi_1 - \Phi_0) - P_2(\Phi_2 - \Phi_0) + P_1 Q_1 + P_2 Q_2 \\ P_3(\Phi_3 - \Phi_0) - P_4(\Phi_4 - \Phi_0) + P_3 Q_3 + P_4 Q_4 \\ P_5(\Phi_5 - \Phi_0) - P_6(\Phi_6 - \Phi_0) + P_5 Q_5 + P_6 Q_6 \end{pmatrix} \quad (29)
$$

2. Add all rows of Eq. (27) and use the fact that the solution obeys Laplace's equation: $m(\Phi_{xx} + \Phi_{yy} + \Phi_{zz}) = 0$, where m is any number. Here we divide by 2 so that the final coefficients will be in the standard form (i.e. $m = 1$):

$$
\frac{1}{2}[P_1(\Phi_1 - \Phi_0) + P_2(\Phi_2 - \Phi_0) + P_3(\Phi_3 - \Phi_0)
$$
$$
+ P_4(\Phi_4 - \Phi_0) + P_5(\Phi_5 - \Phi_0) + P_6(\Phi_6 - \Phi_0)]
$$
$$
= \frac{1}{2}\begin{pmatrix} A_1 & A_2 & A_3 \end{pmatrix}\begin{pmatrix} \Phi_x \\ \Phi_y \\ \Phi_z \end{pmatrix} + \frac{1}{2}A_4 \quad (30)
$$

where

$$
\begin{aligned}
A_1 &= P_1 R_{1,1} - P_2 R_{2,1} + P_3 R_{3,1} - P_4 R_{4,1} + P_5 R_{5,1} - P_6 R_{6,1} \\
A_2 &= P_1 R_{1,2} - P_2 R_{2,2} + P_3 R_{3,2} - P_4 R_{4,2} + P_5 R_{5,2} - P_6 R_{6,2} \\
A_3 &= P_1 R_{1,3} - P_2 R_{2,3} + P_3 R_{3,3} - P_4 R_{4,3} + P_5 R_{5,3} - P_6 R_{6,3} \\
A_4 &= -P_1 Q_1 + P_2 Q_2 - P_3 Q_3 + P_4 Q_4 - P_5 Q_5 + P_6 Q_6 \quad (31)
\end{aligned}
$$

To eliminate the field terms, Φ_x, Φ_y, and Φ_z, Eq. (29) can now be substituted into Eq. (30):

$$
\frac{1}{2}[P_1(\Phi_1 - \Phi_0) + P_2(\Phi_2 - \Phi_0) + P_3(\Phi_3 - \Phi_0)
$$
$$
+ P_4(\Phi_4 - \Phi_0) + P_5(\Phi_5 - \Phi_0) + P_6(\Phi_6 - \Phi_0)]
$$
$$
= \frac{1}{2}\begin{pmatrix} A_1 & A_2 & A_3 \end{pmatrix}\begin{pmatrix} D_{1,1} & D_{1,2} & D_{1,3} \\ D_{2,1} & D_{2,2} & D_{2,3} \\ D_{3,1} & D_{3,2} & D_{3,3} \end{pmatrix}
$$

$$
\times \begin{pmatrix} P_1(\Phi_1 - \Phi_0) - P_2(\Phi_2 - \Phi_0) + P_1 Q_1 + P_2 Q_2 \\ P_3(\Phi_3 - \Phi_0) - P_4(\Phi_4 - \Phi_0) + P_3 Q_3 + P_4 Q_4 \\ P_5(\Phi_5 - \Phi_0) - P_6(\Phi_6 - \Phi_0) + P_5 Q_5 + P_6 Q_6 \end{pmatrix} + \frac{1}{2} A_4
$$

$$
= \frac{1}{2} \begin{pmatrix} W_1 & W_2 & W_3 \end{pmatrix}
$$

$$
\times \begin{pmatrix} P_1(\Phi_1 - \Phi_0) - P_2(\Phi_2 - \Phi_0) + P_1 Q_1 + P_2 Q_2 \\ P_3(\Phi_3 - \Phi_0) - P_4(\Phi_4 - \Phi_0) + P_3 Q_3 + P_4 Q_4 \\ P_5(\Phi_5 - \Phi_0) - P_6(\Phi_6 - \Phi_0) + P_5 Q_5 + P_6 Q_6 \end{pmatrix} + \frac{1}{2} A_4
$$

$$(32)$$

where

$$
\begin{aligned}
W_1 &= A_1 D_{1,1} + A_2 D_{2,1} + A_3 D_{3,1} \\
W_2 &= A_1 D_{1,2} + A_2 D_{2,2} + A_3 D_{3,2} \\
W_3 &= A_1 D_{1,3} + A_2 D_{2,3} + A_3 D_{3,3}
\end{aligned}
$$

$$(33)$$

Multiplying out this equation and substituting for A_4 from Eq. (31), one obtains, after collecting terms in Φ_n ($n = 0 \ldots 6$), an equation relating the potential at the center node, Φ_0 (see Fig. 1), to the potentials at its six nearest neighboring nodes, $\Phi_1 \ldots \Phi_6$, as required, that can be written in the standard finite difference equation form:

$$
\boxed{\alpha_1 \Phi_1 + \alpha_2 \Phi_2 + \alpha_3 \Phi_3 + \alpha_4 \Phi_4 + \alpha_5 \Phi_5 + \alpha_6 \Phi_6 = \alpha_0 \Phi_0 + \alpha_7}
$$

$$(34)$$

where the coefficients $\alpha_0 \ldots \alpha_7$ are given in Table 3.

Comparing Eq. (34) with Eq. (5), one can see that they are almost identical in form. This means that the same solution method that was used for the 3D electrode meshes can still be used to solve for meshes containing charged dielectric structures as well. An additional term, α_7, is present in Eq. (34) and this represents the contribution to the potential distribution from any surface charges present on the dielectrics ($\alpha_7 = 0$ if the dielectric materials are uncharged).

In summary, the finite difference coefficients, $\alpha_0 \ldots \alpha_7$, are calculated as follows. First the coefficients R, $P(= 1/S)$, and Q in Eq. (25) can all be deduced from a knowledge of the mesh, the nature of the surface of the material (and any charges on it), and the position where the surface cuts the mesh arms, using the formulæ in Table 2. Using the P and R coefficients, one can generate the matrix $\mathbf{C}$ (as defined in Eq. (28)) and hence its inverse, $\mathbf{D}$. The P and R coefficients are then used to generate the

Table 3 Formulæ for the coefficients of the final FD equation

$\alpha_1 = \frac{P_1}{2}(1 - W_1)$	$\alpha_2 = \frac{P_2}{2}(1 + W_1)$
$\alpha_3 = \frac{P_3}{2}(1 - W_2)$	$\alpha_4 = \frac{P_4}{2}(1 + W_2)$
$\alpha_5 = \frac{P_5}{2}(1 - W_3)$	$\alpha_6 = \frac{P_6}{2}(1 + W_3)$

$$\alpha_0 = \alpha_1 + \alpha_2 + \alpha_3 + \alpha_4 + \alpha_5 + \alpha_6$$
$$\alpha_7 = -\alpha_1 Q_1 + \alpha_2 Q_2 - \alpha_3 Q_3 + \alpha_4 Q_4 - \alpha_5 Q_5 + \alpha_6 Q_6$$

coefficients A_1, A_2, and A_3 (Eq. (31)). Next, the W coefficients are formed from the D's and A's (Eq. (33)). Finally, using the P, W, and Q variables, the coefficients of the FD equation, $\alpha_1 \ldots \alpha_7$, can be formed using Table 3.

2.3.2 Multiple Dielectric Cuts on a Mesh Arm

In the previous subsection, an FD equation (34) was derived which allowed for a single cut on any one arm of the FD mesh. However, one may ask what should be done if the mesh is so coarse or the structure so fine that two interfaces intersect a single mesh line element? Initially, one may say that if this occurs, then the mesh should be made finer so as to ensure that at least one mesh-point lies between two dielectric interfaces. However, since the grid is rectangular, this becomes expensive when one models a thin dielectric spherical shell, for example, since the whole mesh must be made very fine to satisfy the above requirement of having one node inside the material. It would be preferable to formulate the equations in such a way that multiple dielectric cuts on any one arm can be modeled.

If we aim to form a matrix $\underline{R}$ and vectors $\mathbf{S}$ and $\mathbf{Q}$, relating the nodal potentials, $\Phi_1 \ldots \Phi_6$, to the derivatives of the potential at the center node, Φ_x, Φ_y, Φ_z and Φ_{xx}, Φ_{yy}, Φ_{zz}, in the form of Eq. (26), then we can eliminate the derivatives and form the FD equations exactly as before (Eqs. (27)–(34)). The question therefore becomes; can $\underline{R}$, $\mathbf{S}$, and $\mathbf{Q}$ be formed such that multiple dielectric surface cuts on any one arm can be modeled?

Consider Eq. (23); it can be written in the following way:

$$\Phi_1 = \Phi_0 + h_1 \Phi_x + \frac{1}{2} h_1^2 \Phi_{xx}$$

$$+ b_1^{(1)} \left\{ \gamma_1^{(1)} n_{x1}^{(1)} (n_{x1}^{(1)} \Phi_x + n_{y1}^{(1)} \Phi_y + n_{z1}^{(1)} \Phi_z) + a_1^{(1)} \gamma_1^{(1)} n_{x1}^{(1)2} \Phi_{xx} - \frac{\sigma_1^{(1)} n_{x1}^{(1)}}{\epsilon_0 \epsilon_{r1}^{(1)}} \right\}$$

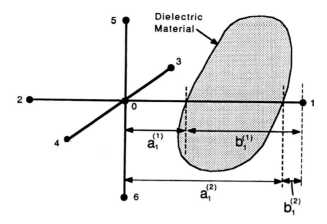

Figure 7 Case of two dielectric surfaces cutting mesh arm number 1.

where the superscript, (1), denotes the **first** cut on the arm. The subscript 1 means **arm number 1**, as before. This can be written as:

$$\Phi_1 = \Phi_0 + h_1 \Phi_x + \frac{1}{2} h_1^2 \Phi_{xx} + b_1^{(1)} \Delta \Phi_{x1}^{(1)} \tag{35}$$

where

$$\Delta \Phi_{x1}^{(1)} = \gamma_1^{(1)} n_{x1}^{(1)} (n_{x1}^{(1)} \Phi_x + n_{y1}^{(1)} \Phi_y + n_{z1}^{(1)} \Phi_z) + a_1^{(1)} \gamma_1^{(1)} n_{x1}^{(1)^2} \Phi_{xx} - \frac{\sigma_1^{(1)} n_{x1}^{(1)}}{\epsilon_0 \epsilon_{r1}^{(1)}} \tag{36}$$

Thus the potential, Φ_1, at the end of arm 1 is given by the standard Taylor expansion of Φ about 0: $\Phi_0 + h_1 \Phi_x + \frac{1}{2} h_1^2 \Phi_{xx}$, plus an additional term, $b_1^{(1)} \Delta \Phi_{x1}^{(1)}$, which accounts for the crossing of the first dielectric cut.

Suppose there is a second dielectric cut on arm 1 at a distance $a_1^{(2)}$ from the center node, 0, and $b_1^{(2)}$ from node 1 (see Fig. 7). At this cut, by using the same Taylor series expansion method and interface conditions as described in the previous subsection, the potential, Φ_1, at the end of arm 1 can be written in the form:

$$\Phi_1 = \Phi_0 + h_1 \Phi_x + \frac{1}{2} h_1^2 \Phi_{xx} + b_1^{(1)} \Delta \Phi_{x1}^{(1)} + b_1^{(2)} \Delta \Phi_{x1}^{(2)} \tag{37}$$

$b_1^{(1)} \Delta \Phi_{x1}^{(1)}$ represents the change in the potential Φ due to the field discontinuity across the first cut, and $b_1^{(2)} \Delta \Phi_{x1}^{(2)}$ represents the corresponding effect resulting from the field discontinuity at the second cut. By analogy with Eq. (36), the expression for $\Delta\Phi_{x1}^{(2)}$ is given by:

$$\Delta\Phi_{x1}^{(2)} = \gamma_1^{(2)} n_{x1}^{(2)} \left(n_{x1}^{(2)} \Phi_x + n_{y1}^{(2)} \Phi_y + n_{z1}^{(2)} \Phi_z \right) + a_1^{(2)} \gamma_1^{(2)} n_{x1}^{(2)^2} \Phi_{xx} - \frac{\sigma_1^{(2)} n_{x1}^{(2)}}{\epsilon_0 \epsilon_{r1}^{(2)}} \quad (38)$$

Substituting Eqs. (36) and (38) into Eq. (37) and rearranging, gives modified formulæ for the coefficients $R_{i,j}$, S_i, and Q_i in Table 2. For the case of two cuts on arm 1, the resulting formulæ are:

$$R_{1,1} = h_1 + b_1^{(1)} \gamma_1^{(1)} n_{x1}^{(1)^2} + b_1^{(2)} \gamma_1^{(2)} n_{x1}^{(2)^2}$$

$$R_{2,1} = b_1^{(1)} \gamma_1^{(1)} n_{x1}^{(1)} n_{y1}^{(1)} + b_1^{(2)} \gamma_1^{(2)} n_{x1}^{(2)} n_{y1}^{(2)}$$

$$R_{3,1} = b_1^{(1)} \gamma_1^{(1)} n_{x1}^{(1)} n_{z1}^{(1)} + b_1^{(2)} \gamma_1^{(2)} n_{x1}^{(2)} n_{z1}^{(2)}$$

$$S_1 = \frac{1}{2} h_1^2 + a_1^{(1)} b_1^{(1)} \gamma_1^{(1)} n_{x1}^{(1)^2} + a_1^{(2)} b_1^{(2)} \gamma_1^{(2)} n_{x1}^{(2)^2}$$

$$Q_1 = \frac{b_1^{(1)} \sigma_1^{(1)} n_{x1}^{(1)}}{\epsilon_0 \epsilon_{r1}^{(1)}} + \frac{b_1^{(2)} \sigma_1^{(2)} n_{x1}^{(2)}}{\epsilon_0 \epsilon_{r1}^{(2)}}$$

Similar formulæ are obtained for two cuts on each of the other five arms.

For m dielectric cuts on arm number 1, the potential at the end of the arm, Φ_1, can be written:

$$\Phi_1 = \Phi_0 + h_1 \Phi_x + \frac{1}{2} h_1^2 \Phi_{xx} + b_1^{(1)} \Delta\Phi_{x1}^{(1)} + b_1^{(2)} \Delta\Phi_{x1}^{(2)} + \cdots + b_1^{(m)} \Delta\Phi_{x1}^{(m)} \quad (39)$$

where $b_1^{(i)} \Delta\Phi_{x1}^{(i)}$ ($i = 1 \ldots m$), represents the change in potential due to the field jump across each dielectric cut.

This modified formulation enables multiple dielectric cuts on a single arm to be handled. In practice, it was found that a less accurate result was obtained when two dielectric surfaces cut a single mesh line. Therefore, wherever possible, at least one mesh node should be kept between two dielectric surfaces, especially if the surfaces form part of a region of particular interest.

2.3.3 Magnetic Formulation

We also want to handle magnetic field computations in 3D for systems including ferromagnetic materials and coil windings. The conventional method of doing this is to use a vector potential $\mathbf{A}$ (Munro, 1973). For 2D problems, such as rotationally symmetric magnetic lenses, this approach works very well, since the field can be fully described using a single component of $\mathbf{A}$, e.g., A_θ. In 3D, unfortunately, this simplification is impossible, and three components (A_x, A_y, A_z) would be required at each node. For 3D problems, therefore, a different formulation proposed by Simpkin and Trowbridge (1979) is preferable, using a magnetic scalar potential, Φ_m, which requires only one function value per node to be stored.

To do this, we express the total magnetic field at any point, $\mathbf{H}$, as the sum of the field due to the coils, $\mathbf{H}_c$, plus the field due to the magnetization of the iron, $\mathbf{H}_m$:

$$\mathbf{H} = \mathbf{H}_c + \mathbf{H}_m \tag{40}$$

Now, from Maxwell's equations for static magnetic fields, $\nabla \wedge \mathbf{H} = \mathbf{J}$, and hence

$$\nabla \wedge \mathbf{H}_c + \nabla \wedge \mathbf{H}_m = \mathbf{J} \tag{41}$$

But the coil field $\mathbf{H}_c$ also satisfies $\nabla \wedge \mathbf{H}_c = \mathbf{J}$, and hence from Eq. (41) it follows that $\nabla \wedge \mathbf{H}_m = \mathbf{0}$. Therefore $\mathbf{H}_m$ can be expressed in terms of a magnetic scalar potential Φ_m thus,

$$\mathbf{H}_m = -\nabla \Phi_m \tag{42}$$

From Eq. (40), therefore, the total field $\mathbf{H}$ is given by

$$\mathbf{H} = \mathbf{H}_c - \nabla \Phi_m \tag{43}$$

The coil field $\mathbf{H}_c$ can be computed, for any set of coil windings, using the Biot–Savart law:

$$H_c = \oint \frac{I\mathbf{dl} \wedge \hat{\mathbf{r}}}{4\pi r^2} \tag{44}$$

which can be evaluated by numerical integration. Thus the total field $\mathbf{H}$ can be computed if we can evaluate Φ_m at each point on the FD mesh.

As in the electrostatic case, the form of the FD equation depends on the locations of the nodes. If all seven nodes under consideration lie in a region of constant permeability, μ, the Maxwell equation $\nabla . \mathbf{B} = 0$ can be simplified to $\nabla . \mathbf{H} = 0$, and since also $\nabla . \mathbf{H}_c = 0$, it follows from Eq. (42) that in this case

$$\nabla^2 \Phi_m = 0 \tag{45}$$

i.e., Φ_m obeys Laplace's equation for this case and, therefore, the coefficients, α, of the FD equation are simply those given for Eq. (34).

If, however, the central node, 0, is adjacent to a magnetic material, such that one or more of the arms is cut by an interface between two media, of relative permeability μ_{r0} and μ_{r1} say, then Laplace's equation (45) no longer holds. Instead, we must use the interface condition that the normal component of **B** is continuous, i.e.

$$\mu_0 \mu_{r1} H_{n1} = \mu_0 \mu_{r0} H_{n0}$$

Note that μ_0 is the magnetic constant ($4\pi \times 10^{-7}$ H/m) and μ_{r0} is the relative permeability at the center node.

Writing this in terms of Eq. (40):

$$\mu_{r1} (H_{cn1} + H_{mn1}) = \mu_{r0} (H_{cn0} + H_{mn0})$$

thus

$$\mu_{r1} H_{mn1} - \mu_{r0} H_{mn0} = (\mu_{r0} - \mu_{r1}) H_{cn} \tag{46}$$

where H_{cn} is the component of the coil field normal to the interface, (which is the same on both sides of the interface), H_{mn0} is the normal component of the magnetization field on the μ_{r0} side of the interface and H_{mn1} is the normal component of the magnetization field on the μ_{r1} side of the interface. Comparing this equation with its electrostatic analogue (Eq. (13)), we see that the equations are equivalent if we define an effective magnetic surface charge density, σ_m, given by

$$\sigma_m = (\mu_{r0} - \mu_{r1}) H_{cn} \tag{47}$$

The procedure, therefore, is simply to compute H_{cn} with the Biot–Savart law wherever a magnetic interface cuts the FD mesh, and then use the value of σ_m given by Eq. (47) to set up the corresponding FD equation, exactly as for the electrostatic case. This is, therefore, an extremely convenient method for computing the magnetic distribution, since it requires minimal changes to the existing electrostatic program (mainly, the addition of routines to compute the Biot–Savart coil field for different coil arrangements).

3. IMPLEMENTATION OF TWO NEW SOFTWARE PACKAGES

3.1 Introduction

Two distinct software packages have been developed, based on the theory of the preceding section for electrostatic and magnetic systems respectively. The packages are named ELEC3D for the electrostatic analyses and MAG3D for magnetic analyses. In addition, a program has been written which performs direct electron ray-tracing in 3D systems containing regions of overlapping electric and magnetic fields. An annotated list of programs contained in ELEC3D is given below; the programs in the MAG3D package are functionally similar, with any differences mentioned in parentheses:

i.   **EPREVIEW** – allows 2D sections of the 3D finite difference mesh to be previewed with electrode (polepiece) geometry superposed.

ii. **ESETUP** – a preprocessor program which sets up the finite difference equations and boundary conditions.

iii. **ESOLVE** – solves the finite difference equations, outputting the electrostatic potential (magnetic scalar potentials) at each mesh point.

iv. **EAXIAL** – Extracts and outputs potentials along a specified straight axis through the 3D mesh. The program also computes various axial field functions, such as round lens fields, deflection fields, and quadrupole fields. The axial potentials can also be output in a form readable by the Imperial College Lithography (Chu & Munro, 1981) and Stray Field (Zhu, 1989) electron optical design packages.[1]

v.   **ETRAJ** – computes electron trajectories through 3D electric (magnetic) fields.

vi. **ECONT** – plots equipotentials of electrostatic potentials (magnetic scalar potentials) in 2D sections of the 3D mesh, with a superposition of the electrode (polepiece) geometry. The program also plots 2D projections of the trajectories computed by program ETRAJ.

Fig. 8 is a graphical representation of how the individual programs in the packages are used together to analyze a 3D problem.

3.2 Representation of the Objects

Finding the optimum way to describe fully 3D objects for input data to a program is a problem which requires more thought than may initially be imagined. (The same is also true for the representation of the 3D results.) In two dimensions, one may draw an object on a flat piece of paper, assuming the surface of the solid object to be an extrusion or rotation about an axis of the set of one dimensional lines and curves. This data is readily input to

---

[1] These pieces of software compute the optical properties of combinations of magnetic and electrostatic lenses and deflectors and the optical effects of non-rotationally symmetric or stray fields in electron optical columns.

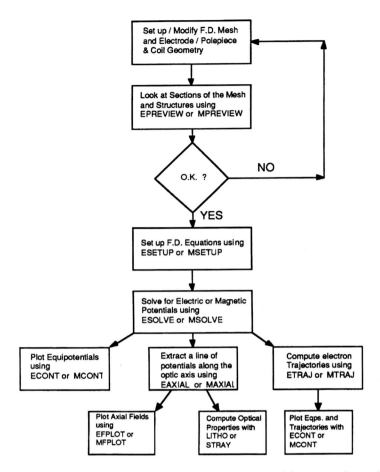

Figure 8 Flowchart showing the scheme for analyzing a 3D problem using the packages ELEC3D and MAG3D.

the computer and the process can be further aided by the use of a digitizer to transfer key points into the computer. In 3D, one must consider the extent and relative positions of various objects in a third direction, so if complicated and general 3D objects are to be analyzed (i.e., not just those which are surfaces of revolution or extrusions of 2D objects), then one must choose a correspondingly general way of describing their position and shape.

The method used in these programs is to define the volume of each object as that bounded by the intersection of any number of quadric surfaces. A quadric function is a second degree polynomial in x, y, z and may

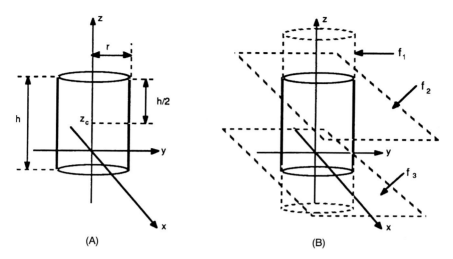

Figure 9 Illustration of how a 3D solid object is represented by a set of intersecting surfaces for the case of a solid cylinder with its axis along the z-axis: (A) geometry; (B) construction using three intersecting quadric surfaces.

be written in the general form:

$$f(x, y, z) = C_0 + C_1 x + C_2 y + C_3 z + C_4 x^2 + C_5 y^2 + C_6 z^2 + C_7 xy$$
$$+ C_8 xz + C_9 yz \tag{48}$$

The convention used in the software is that $f(x, y, z) = 0$ defines the surface (or part of the surface) of a 3D object, $f(x, y, z) > 0$ is inside the object, and $f(x, y, z) < 0$ is outside the object. The surfaces of many common 3D objects (e.g., spheres, cylinders, cones, boxes) can be described by quadric equations. If the bounding surface of an object is defined by several of these surfaces $(f_1(x, y, z), f_2(x, y, z), \ldots)$, then if, at any point (X, Y, Z), the minimum of $(f_1(X, Y, Z), f_2(X, Y, Z), \ldots)$ is negative, the point is outside the object. By evaluating this minimum of the function values for any object at all the mesh points, one can map the physical extent of the object.

As an example, consider a solid cylinder which lies along the z-axis with a height $h = 5$ mm, radius $r = 3$ mm, and is centered at $z_c = 2$ mm (see Fig. 9A). This object may be described using three intersecting quadric surfaces (Fig. 9B).

The curved surface of the cylinder, f_1, is defined thus:

$$f_1(x, y, z) = 9 - x^2 - y^2 \tag{49}$$

[Notice that the equation is **not** defined as $f_1(x, y, z) = -9 + x^2 + y^2$. This would mean that all points at a radius greater than 3 mm would have a positive function value (i.e., by definition would be inside the object) and all points at a radius less than 3 mm would have a negative function value (i.e., by definition would be outside the object). In this case the function would describe a cylindrically hollow void – this is useful, however, if a hole has to be drilled in a solid object.]

The other two surfaces define the extent of the object in the z-direction. The first is a plane surface, f_2, which defines all points for which $z < -0.5$ mm to be outside the object ($z_c - h/2 = -0.5$ mm):

$$f_2(x, y, z) = 0.5 + z \tag{50}$$

The second is a plane surface, f_3, which defines all points for which $z > 4.5$ mm to be outside the object ($z_c + h/2 = 4.5$):

$$f_3(x, y, z) = 4.5 - z \tag{51}$$

If we define a function $g(x, y, z)$ where:

$$g(x, y, z) = MINIMUM(\, f_1(x, y, z), \ f_2(x, y, z), \ f_3(x, y, z) \,) \tag{52}$$

then $g(x, y, z) = 0$ defines the surface of the solid cylinder, $g(x, y, z) > 0$ defines the interior of the cylinder, and $g(x, y, z) < 0$ defines the space outside the cylinder.

In this way, many 3D objects can be assembled as a set of intersecting quadric surfaces. A useful feature of this method is that holes can be cut through previously defined objects, simply by defining an interior surface. For simple objects it is tedious to keep having to define an object as a set of surfaces, so a few commonly used shapes can be defined by a single line in the data. These are then broken down into their constituent quadric surfaces by a pre-processing subroutine in the program. In principle, many complicated objects could be defined by a single definition if a suitable pre-processor is written to break the object down into quadric surfaces. The object descriptions currently accepted by the pre-processor are listed in Table 4 and illustrated in Fig. 10.

3.3 Setting up the Equations – Programs ESETUP and MSETUP

The task of computing the potential distribution from the input data is divided between two programs, ESETUP and ESOLVE (or MSETUP and

Table 4 Command list for object definition accepted by the pre-processor

CYLINDER	$x_1, y_1, z_1, x_2, y_2, z_2, r_1, r_2$	Defines a cylinder
SPHERE	x_c, y_c, z_c, r_1, r_2	Defines a sphere
BOX	$x_c, y_c, z_c, x_p, y_p, z_p, x_q, y_q, z_q, s_u, s_v, s_w$	Defines a box
CONE	$x_1, y_1, z_1, x_2, y_2, z_2, \alpha$	Defines a cone
FRUSTRUM	$x_1, y_1, z_1, x_2, y_2, z_2, r_1, r_2, r_3, r_4$	Defines a frustrum
CYLHOLE	$x_1, y_1, z_1, x_2, y_2, z_2, r$	Defines a cylindrical hole
SPHHOLE	x_c, y_c, z_c, r	Defines a spherical hole
CONHOLE	$x_1, y_1, z_1, x_2, y_2, z_2, \alpha$	Defines a conical hole
PLANE	$x_p, y_p, z_p, x_n, y_n, z_n$	Defines a plane cut
GENERAL	$c_0, c_1, c_2, c_3, c_4, c_5, c_6, c_7, c_8, c_9$	Defines a general quadric surface

MSOLVE for magnetic analyses). The program ESETUP is essentially a pre-processor program which decodes the input data and formulates the finite difference equations and the boundary conditions. The program ESOLVE reads in this prepared data and solves the equations to find the potential distribution. The reason for splitting the job in two is that it allows the more memory intensive and time consuming part (i.e., the solver, ESOLVE) to be reasonably compact and dedicated to simply solving the equations. This is especially true with the FDM, since the formulation of the equations is quite a logically and memory intensive step in its own right. (As mentioned previously, the advantage for the user of decoupling the mesh generation from the definition of the objects has the disadvantage for the programmer that it requires more effort to interpret the data and set up the equations.) This section looks at this equation set-up program, ESETUP, and describes the main tasks that it performs in order to compute the FD equations in a fast and memory efficient manner.

Fig. 11 is a breakdown of the ESETUP or MSETUP program into its main tasks. Firstly, the input data must be analyzed. This involves setting up the finite difference mesh, breaking down the objects into their constituent quadric surfaces and, in the case of a magnetic analysis, setting up the coils.

A typical data file is given in Fig. 12, corresponding to a magnetic lens consisting of four cylindrical objects. The first three sections of data specify the coordinates of the mesh lines along the X-, Y-, and Z-axes, in that order. The general format of the mesh specification is shown in Fig. 13.

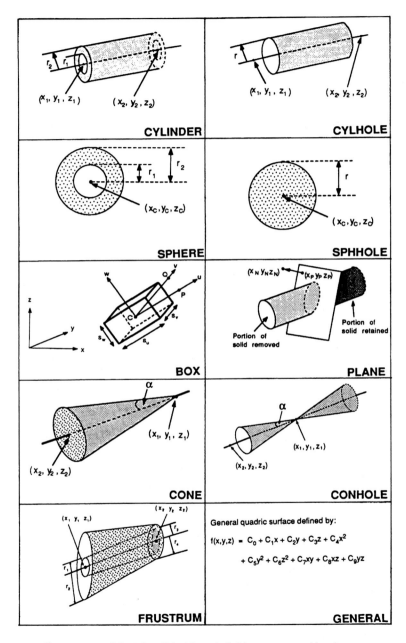

Figure 10 Illustrations of the 3D solid object definitions accepted by the pre-processor.

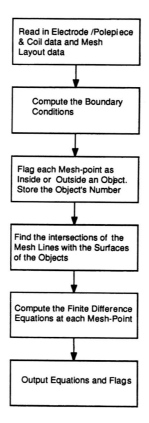

Figure 11 Flowchart showing the tasks performed by ESETUP or MSETUP for setting up the finite difference equations.

Each axis has a title *-COORDINATES in upper case (where * is X, Y, or Z). The position, p_n of the nth principal mesh line is then specified for $1 \leq n \leq NMAX$, where $NMAX$ is specified in a parameter statement in the program. (The values currently set for $NMAX$ are 200 for 8 Mbytes RAM, 100 for 4 Mbytes, and 50 for 2 Mbytes.) In double precision with 8 Mbytes of RAM the maximum number of nodes the mesh can contain has been set at 200,000 (104,000 for 4 Mbytes, 40,000 for 2 Mbytes). The number of nodes along each axis is $i_n + 2$, $j_n + 2$, and $k_n + 2$. (Two extra nodes on each axis are used by the program, in addition to those specified by the user. These extra nodes are for storing the boundary conditions, as will be described later in this section.) Therefore, in 8 Mbytes, the mesh must be arranged so that $(i_n + 2) \times (j_n + 2) \times (k_n + 2) \leq 200,000$.

```
X-COORDINATES
        1       -150.00000
       14        -49.910000
       17        -25.019000
       25          0.000000
       33         25.019000
       36         49.910000
       49        150.00000
/
Y-COORDINATES
        1       -150.00000
       14        -49.910000
       17        -25.019000
       25          0.000000
       33         25.019000
       36         49.910000
       49        150.00000
/
Z-COORDINATES
        1       -200.00000
        4       -150.01000
        6       -119.99000
       16        -25.00100
       26         25.00100
       30         40.01000
       40        100.00000
/

POLEPIECE   1000
CYLINDER  0, 0,    25,    0, 0, 40,   10, 90
BOUNDS       -108   108  -108   108           20   50

POLEPIECE   1000
CYLINDER  0, 0,  -150,    0, 0, 40,   90   115
BOUNDS       -158   158  -158   158          -200   50

POLEPIECE   1000
CYLINDER  0, 0,  -150,    0, 0, -120,  40, 90
BOUNDS       -108   108  -108   108          -200  -100

POLEPIECE   1000
CYLINDER  0, C,  -120,    0, 0, -25,   40, 65
BOUNDS       -108   108  -108   108          -150    0

ROUND COIL   0, 0, -110   0, 0, 0     70  .166666666
ROUND COIL   0, 0,  -75   0, 0, 0     70  .166666666
ROUND COIL   0, 0,  -35   0, 0, 0     70  .166666666

ROUND COIL   0, 0, -110   0, 0, 0     85  .166666666
ROUND COIL   0, 0,  -75   0, 0, 0     85  .166666666
ROUND COIL   0, 0,  -35   0, 0, 0     85  .166666666

BOUNDARIES 2 2 2 2 2
```

Figure 12 Typical input data file for a magnetic lens showing the relative simplicity of the 3D data specification and the independence of the mesh layout from the object definitions.

The lines between the principal mesh lines are automatically equally spaced. When the position of each principal mesh line for an axis has been defined, the data is terminated with a '/'. When this has been completed for each of the three axes, the finite difference mesh layout has been specified.

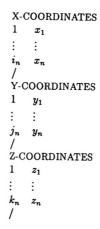

Figure 13 Format of the mesh specification section of the input data.

Each structure is defined as POLEPIECE (in upper case characters) and its relative permeability is also specified on the first line. (For example, a polepiece with relative permeability 1000 would be defined as POLE-PIECE 1000, as shown in the sample data of Fig. 12.)

In the electrostatic package, each structure is defined as either ELEC-TRODE, GRID, or DIELECTRIC (in upper case characters) and its potential (in volts) or relative permittivity is also specified on the first line. (For example, an electrode at 100 volts would be specified as ELEC-TRODE 100, and a dielectric with relative permittivity 3 would be defined as DIELECTRIC 3.) A GRID is treated in an identical way to an ELEC-TRODE for the potential computation, but GRIDS are transparent to the electrons in the ray-trace.

The subsequent data lines specify the mathematical functions that define the surfaces of each object. This is done either by specifying the coefficients of a quadratic polynomial in x, y, z (called a GENERAL surface), or by using key-words for the several common geometrical shapes shown in Fig. 10 and listed in Table 4. In this example (Fig. 12), they are all cylindrical objects defined by the key-word CYLINDER, but could, in principle, be any of the shapes listed in Table 4. In the later sections of the chapter, examples are given where more varied shapes are used.

After each of the object surfaces have been defined, the bounds of the object in 3D space are then defined using the command BOUNDS. These bounding values may be set to the boundary values of the mesh, but the speed of the programs can be greatly enhanced if the approximate region in space occupied by a particular object is known by the program. These maximum and minimum values on x, y, and z must completely surround the object and it is a good idea to put the object in a slightly bigger "*box*" than is necessary (unless one side of the object is on a mesh boundary, in which case the value of that boundary must be given as the bound on the object's volume).

Circular coils are specified by the ROUND COIL commands. Straight, current carrying wires can also be defined in the data by the STRAIGHT WIRE command.

When all the objects and coils have been defined, the boundary conditions at each face of the mesh are defined with the command BOUNDARIES. The form of the command is:

BOUNDARIES IBX1, IBX2 IBY1, IBY2 IBZ1, IBZ2

where:

$$IBX1 = \text{lower } x \text{ boundary type}$$
$$IBX2 = \text{upper } x \text{ boundary type}$$
$$IBY1 = \text{lower } y \text{ boundary type}$$
$$IBY2 = \text{upper } y \text{ boundary type}$$
$$IBZ1 = \text{lower } z \text{ boundary type}$$
$$IBZ2 = \text{upper } z \text{ boundary type}$$

In the magnetic program, the values of these integers can be as follows:

$1 =$ Neumann boundary for the magnetic scalar potentials ($\partial \Phi_m / \partial n = 0$)
$2 =$ Zero field boundary, where the total normal component of **H** is zero
$3 =$ Dirichlet boundary, where the magnetic scalar potential is zero

and in the electrostatic program, the values of these integers can be as follows:

$1 =$ Neumann boundary for the electrostatic potentials ($\partial \Phi / \partial n = 0$)
$3 =$ Dirichlet boundary, where the electrostatic potential is zero

In general, as Fig. 12 illustrates, the mesh specification is decoupled from the specification of the lens. The only thing the user must ensure is that there are enough mesh points in certain regions (e.g., in the lens gap, where the field gradient is large), to enable an accurate result to be computed. Fig. 14 is a section through this mesh and lens structure obtained with the program MPREVIEW.

Once the program has decoded the input data, it then sets up the boundary conditions. Outside each of the six faces of the mesh there is

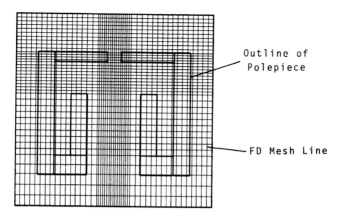

Figure 14 Section through the illustrative magnetic lens obtained using program MPREVIEW, showing the polepiece geometry and the finite difference mesh superposed.

Figure 15 One dimensional line of potentials illustrating the use of auxiliary border potentials to apply the boundary conditions.

an auxiliary plane of potentials. If Neumann boundary conditions are to be applied (i.e., $\partial\Phi/\partial n = 0$, where n is the normal to the boundary surface), then this outer, auxiliary plane of potentials contains a direct copy of the potentials in the penultimate plane of the mesh. Considering the 1D analogue of this (Fig. 15), suppose there is a line of N potentials, Φ_n ($n = 1, 2, \ldots, N$). Two extra potentials at $n = 0$ and $n = N + 1$ are added to the line and are set such that $\Phi_0 = \Phi_2$ and $\Phi_{N+1} = \Phi_{N-1}$.

In its discretized form, the Neumann condition at mesh point N is represented by:

$$\frac{\partial\Phi}{\partial n} = \frac{\Phi_{N+1} - \Phi_{N-1}}{2h} = 0 \tag{53}$$

i.e.,

$$\Phi_{N+1} = \Phi_{N-1} \tag{54}$$

where h is the mesh spacing at the edge.

Therefore, by including these extra potentials at the end of each line of the mesh and setting them equal to the penultimate potentials, we automatically ensure that the Neumann condition is obeyed on the boundaries.

Applying the boundary conditions in this way means that the solver applies the finite difference equation at all interior nodes of the mesh, using the same logic for each (i.e., that the new potential is a weighted sum of its neighbors). At a point on the edge of the mesh, one or more of the potentials may be retrieved from the additional border planes, but this is arranged so that it is done without a break in logic in the solver routine. The same is true in the direct ray-trace program, where the potentials surrounding the electron are differenced to obtain the field. Therefore, the solver and ray-trace routines do not have to check if the potential they require is on the edge of the mesh, and if it is, which edge, and then decide what potential to use. (Having to do such checks at each point would slow the routines down considerably and would have meant using many conditional jumps, which would also make the code difficult to follow.)

If Dirichlet boundaries are required (i.e., $\Phi = 0$), then the potentials along the edge of the mesh are fixed to zero and are not relaxed.

In the magnetic program, MSETUP, there is an option to set the total normal field to zero on the boundary. (The $\Phi = 0$ condition just sets the magnetization potentials to zero.) The total normal field component, H_n, is the sum of the normal coil field, H_{cn}, and the normal magnetization field, $-\partial\Phi/\partial n$, at the boundary, i.e.,

$$H_n = H_{cn} - \frac{\partial\Phi}{\partial n}$$

If H_n is to be set to zero, then from Eq. (53), in discretized form:

$$H_{cn} - \frac{\Phi_{N+1} - \Phi_{N-1}}{2h} = 0 \tag{55}$$

where h is the mesh spacing at the edge. Therefore,

$$\Phi_{N+1} = \Phi_{N-1} + 2hH_{cn} \tag{56}$$

i.e., the auxiliary border potential is set to the penultimate potential plus $2hH_{cn}$.

The next step in the setup procedure is to flag each mesh point as to whether it is inside or outside an object. For each mesh point, this is done by finding the minimum of all the quadric functions defining each object at the point under consideration, and assigning the node to an object if the minimum is positive for that object. Points inside an electrode then become dead points, in the sense that they are set at the electrode voltage and are not relaxed. This can speed up the solver, especially if there is a large volume of electrode regions. Points inside dielectrics or magnetic materials are flagged so that their permittivity or permeability can be accessed.

The program then finds the coordinates at which the object surfaces intersect with the mesh. The x-lines, y-lines, and z-lines of the mesh are considered individually. Finding the intersection points of one of these lines with the surfaces is fairly trivial, but some thought must be given as to how these are sorted and stored so that they may be accessed quickly and occupy the minimum memory.

Firstly, all the x-lines are considered. There will be $NY \times NZ$ of these, where NY is the number of mesh nodes in y, and NZ is the number of nodes in z. For any x-line, all the intersections it has with each surface are found and are sorted into ascending order. Then, all physically irrelevant intersections are eliminated. For example, if two dielectrics of the same permittivity are touching, then the program will detect two cuts on the mesh at this point. In reality, there is no physical discontinuity in the material, and so no interface equation is required, so the cuts can be disregarded. (Also, this procedure can detect errors in the input data if, for example, two electrodes of different potentials are touching.) This procedure is repeated for all the y- and z-lines.

All the coordinates of the cuts and a flag identifying the object which has cut the mesh are stored in two master arrays. The contents of these arrays are accessed by the use of three pointing vectors (one for each set of mesh lines), which store the master array addresses of all the cuts on any x-, y-, or z-line.

The next step is to compute the FD equation coefficients. The points are considered in the same order in which they are to be relaxed. In this way, the solver does not have to search for the various parameters it needs in large data arrays – instead it can simply read them off in the same order in which they were written. At each mesh point, the array of surface cut positions is interrogated to deduce the six arm lengths of the node; the object flag array is also used to identify the surfaces (if any) which intersect the mesh near the node. From a knowledge of the surfaces, the normals to them are computed and the relative permittivity of each region is found. Any surface charges at the cut can either be read from the data or computed (in the magnetic case) from the coil geometry via the Biot–Savart law (Eq. (44)). With this information, the coefficients, $\alpha_1 \ldots \alpha_7$, of the finite difference equation (34) of Section 2 are generated, using the formulæ given in Table 3. The mesh flag for each node is coded to tell the solver how many of these coefficients are non-standard (i.e., cannot be computed from the mesh spacings in the vicinity alone). In this way, only these non-standard coefficients need to be stored, and therefore, memory can be saved since there are usually far less non-standard points (i.e., those near the surface of an object) than standard ones.

3.4 Solving the FD Equations – Programs ESOLVE and MSOLVE

The solver programs use the method of over-relaxation, as described in the previous section, to find the potential distribution. The main routine decodes the pointer arrays set up by ESETUP or MSETUP to retrieve any special coefficients for points near surface interfaces. For non-special points, the finite difference equations can be computed quickly from the mesh spacing in the vicinity of the point, using the formulæ given in Table 1 (see Section 2).

The equations are solved in a "chess-board" pattern to ensure that all the potentials used in any equation are in the same iterative state. To see this, consider a 2D example (Fig. 16). The new approximation to the potential at any point is computed as essentially a weighted sum of the potentials at the nearest neighboring nodes, and so if all the potentials at the black dots (Fig. 16) are computed first, then it can be seen that only those potentials occupying the white dots will be used. When all the black-point potentials have been computed, the white-point ones are then calculated using only the black-point potentials just computed (and are, therefore all at the same state of iteration). In the 3D case, of course, the black and white dots are interchanged on each consecutive plane in the third dimension. By the use of the border potential method discussed in the previous section, the boundary conditions are applied with no logical interruptions in the flow of the solution. At the end of each half relaxation pass through the mesh (i.e., all the black points or all the white points), these border potentials are updated to the values in the penultimate planes, before the start of the next half pass through the opposite color points.

The potential distribution is deemed to be solved when none of the nodal potentials are changed on two successive passes by more than a pre-set value defined by the user.

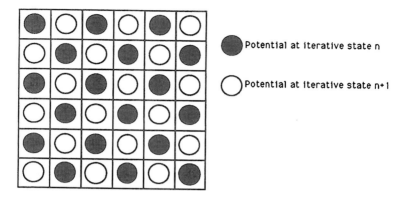

Figure 16 Two dimensional illustration of the chess-board order in which the potentials are relaxed during the solution of the FD equations.

3.5 Direct Electron Ray-Tracing – Programs ETRAJ and MTRAJ

One advantage of computing the potentials on a rectangular grid of points is that it is then a fairly quick job to obtain the fields and higher derivatives of the potentials at the mesh nodes. The first derivatives of the potential $(\partial\Phi/\partial x, \partial\Phi/\partial y, \partial\Phi/\partial z)$ at the nodes are found by taking a three point centered difference in each of the three directions. The second mixed derivatives $(\partial^2\Phi/\partial x\partial y, \partial^2\Phi/\partial y\partial z, \partial^2\Phi/\partial z\partial x)$ are found by differencing the eight surrounding potentials in the two directions required.

For example (see Fig. 17), at point 0,

$$\frac{\partial\Phi}{\partial x} = \frac{h_2^2\Phi_1 - h_1^2\Phi_2 - (h_2^2 - h_1^2)\Phi_0}{h_1 h_2(h_1 + h_2)} \tag{57}$$

and

$$\frac{\partial^2\Phi}{\partial x\partial y} = [\Phi_8 h_2 h_4(h_3 h_2 + h_1 h_4) - \Phi_7 h_2 h_3(h_2 h_4 + h_1 h_3)$$
$$+ \Phi_6 h_1 h_3(h_1 h_4 + h_2 h_3) - \Phi_5 h_1 h_4(h_1 h_3 + h_2 h_4)$$
$$- (h_4^2 - h_3^2)h_1 h_2(\Phi_1 - \Phi_2) - (h_2^2 - h_1^2)h_3 h_4(\Phi_3 - \Phi_4)]$$
$$/[2h_1 h_2 h_3 h_4(h_1 + h_2)(h_3 + h_4)] \tag{58}$$

(If the node where the field is required has an object surface intersection along one of its arms, the above formulæ do not give an accurate estimate of the derivatives. In this case, a more complete analysis should include the foreshortened arm lengths. This has not been considered in the present

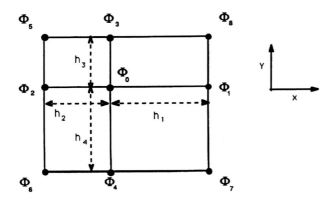

Figure 17 2D grid of potentials and mesh spacings used to form the derivatives for the field computation.

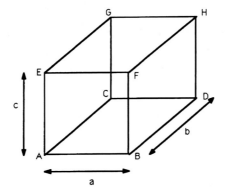

Figure 18 Individual 3D mesh cell with vertices and sides labeled as used in the trajectory computation equation derivations.

programs and this omission may cause some errors in the ray-trace if the electrons pass close to the surface of the objects.)

The third derivative, $\partial^3\Phi/\partial x \partial y \partial z$, can be obtained by differencing the second derivatives in the three different directions, and taking the average of the result, i.e.,

$$\frac{\partial^3\Phi}{\partial x \partial y \partial z} = \frac{1}{3}\left\{ \frac{\partial}{\partial x}\left(\frac{\partial^2\Phi}{\partial y \partial z}\right) + \frac{\partial}{\partial y}\left(\frac{\partial^2\Phi}{\partial z \partial x}\right) + \frac{\partial}{\partial z}\left(\frac{\partial^2\Phi}{\partial x \partial y}\right)\right\} \qquad (59)$$

The mesh can be thought of as a set of 3D boxes, each with eight corner nodes, $A \ldots H$ (see Fig. 18). When an electron enters a specific box, the potential and the seven derivatives, mentioned above, are computed at the eight corners. These 64 values are then used to obtain the potential and first derivatives (x-, y-, z-fields) anywhere inside the box by tri-cubic interpolation.

The potential, $\Phi(x, y, z)$, at any point inside the box can be written as a summation:

$$\Phi(x, y, z) = \sum_{k=0}^{3} \sum_{j=0}^{3} \sum_{i=0}^{3} C_{i,j,k} \left(\frac{x}{a}\right)^i \left(\frac{y}{b}\right)^j \left(\frac{z}{c}\right)^k \tag{60}$$

where a, b, c are the side lengths of the box, and the $C_{i,j,k}$'s are 64 unknowns.

The above polynomial can be differentiated to form the seven derivatives (Φ_x, Φ_y, Φ_z, Φ_{xy}, Φ_{yz}, Φ_{zx}, Φ_{xyz}), and with the potential, Φ, can be used at the eight corners of the mesh, to get 64 equations. The value of the potential and its derivatives are known at the corners of the box and hence the 64 simultaneous equations can be solved to find the $C_{i,j,k}$'s. When these have been determined, the potential and its derivatives can be found at any position (x, y, z) inside the box. It was found that solving the 64 simultaneous equations by Gaussian elimination made the ray-trace very time consuming. The solution was, therefore, programmed explicitly in the code and the ray-trace is now both fast and accurate.

The equations of motion are integrated using a fourth order Runge–Kutta formula (see for example *Numerical Recipes*, by W.H. Press et al., detailed in the book list in the Bibliography) with a variable time step, adjusted so the electron has approximately the same number of steps in each box. On each step, the position of the electron is fed to the minimizing routine of the surface functions, to determine if it has hit any of the objects. The ray-trace is terminated when the electron hits an object or goes out of the mesh.

In the magnetic ray-trace, MTRAJ, the magnetic scalar potentials are differenced to obtain the magnetization field, H_m. The coil field, H_c, is computed by numerical integration of the Biot–Savart equation (44) and added to H_m to obtain the total field.

In the ray-trace through combined 3D electric and magnetic fields, two potential meshes are stored in memory and a separate field computation and interpolation is performed in each to obtain the electric field, $\mathbf{E}$, and magnetic field, $\mathbf{H}$. These are then used in the Lorentz equation to obtain the total force, $\mathbf{F}$, on the electron: $\mathbf{F} = e(\mathbf{E} + \mu_0 \mathbf{v} \wedge \mathbf{H})$, where $\mathbf{v}$ is the electron's velocity, e its charge, and μ_0 the permeability of free space.

3.6 Equipotential Plots – Programs ECONT and MCONT

Plots of electrostatic or magnetic scalar potentials in 2D sections of the mesh can be generated by programs ECONT and MCONT. The sections must be flat layers parallel to the x-, y-, or z-axes. The equipotential plotting program is essentially the same as that described by Munro (1971). A 2D slice of the structure is also superimposed on the plot. This is achieved by evaluating the minimum of all the functions of an object at all the grid points of the selected layer. This gives a set of values ranging from positive (inside the object) to negative (outside the object). The outline of the object can then be displayed by plotting the zero contour on this grid. This is repeated for all the objects.

A 2D projection of the electron trajectories, calculated with the 3D ray-trace programs described in the previous subsection, can also be displayed. Unlike the equipotentials and structures which are slices at the required plane through the mesh, the trajectories are displayed as a view through all the mesh in the required direction. (They are displayed in this way because a 2D slice of a straight line is invariably just a dot; although, in some cases, this may be required, it wasn't considered in the programs thus far.)

Figs. 19 and 20 are two example plots. Fig. 19 is for an electrostatic quadrupole lens triplet with equipotentials and trajectories shown. Fig. 20 is for a magnetic lens with a hole through its side, showing the magnetic scalar equipotentials (Φ_m) and trajectories. The plot of Φ_m shows where the magnetization field ($\nabla\Phi_m$) is strongest (in the gap) but gives no indication of the coil field strength.

3.7 Computing and Plotting Axial Fields – Programs EAXIAL, EFPLOT, MAXIAL, and MFPLOT

The programs EAXIAL and MAXIAL extract electrostatic potentials or magnetic scalar potentials and coil fields from around an optical axis specified by the user. The data from these programs are written to a specified file and may used as input to other programs (such as the STRAYFIELD, Zhu, 1989, or LITHO, Chu & Munro, 1981, suites or to the 3D axial field plotting programs, EFPLOT or MFPLOT). The optical axis can only lie along a line parallel to the x, y, or z axes.

The program computes the axial field function, axial deflection field function and axial quadrupole field functions and outputs them to a specified file. These terms are coefficients in an expansion of the potential near the axis and are defined (for the electrostatic case) in the following derivation.

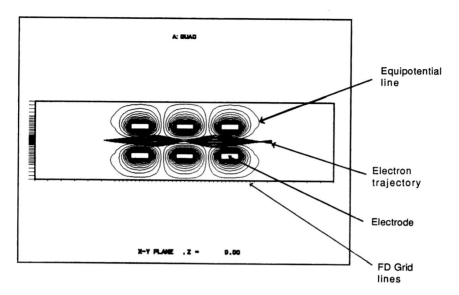

Figure 19 Example of a plot from program ECONT of an electrostatic quadrupole lens showing equipotentials and electron trajectories.

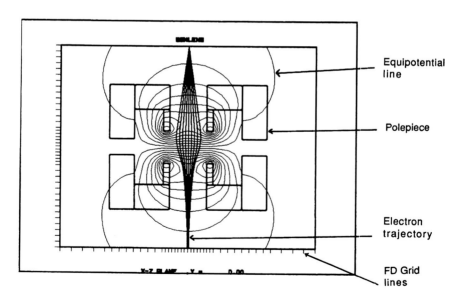

Figure 20 Example of a plot from program MCONT of a round magnetic lens showing magnetic scalar equipotentials and electron trajectories.

Suppose, for example, the optical axis lies along the z-axis. Then the electrostatic potential near the axis can be written as a power series, thus:

$$\Phi(x, y, z) = \Phi_0(z) + a(z)x + b(z)y + c(z)x^2 + d(z)y^2 + e(z)xy$$

Putting $x = r\cos\theta$ and $y = r\sin\theta$, this becomes:

$$\Phi(r, z, \theta) = \Phi_0(z) + a(z)r\cos\theta + b(z)r\sin\theta + \frac{1}{2}c(z)r^2(1 + \cos 2\theta) +$$
$$\frac{1}{2}d(z)r^2(1 - \cos 2\theta) + \frac{1}{2}e(z)r^2 \sin 2\theta$$

or

$$\Phi(r, \theta, z) = \Phi_0(z) + K_1(z)r\cos\theta + K_2(z)r\sin\theta + K_3(z)r^2 \cos 2\theta +$$
$$K_4(z)r^2 \sin 2\theta + K_5(z)r^2$$

where

$\Phi_0(z) =$ axial potential function
$K_1(z) = x$-deflection function
$K_2(z) = y$-deflection function
$K_3(z) =$ first quadrupole function
$K_4(z) =$ second quadrupole function
$K_5(z) =$ round lens focussing function

The programs EFPLOT and MFPLOT plot out the axial field functions generated by the axial field programs. Fig. 21 shows a quadrupole field function plot from the quadrupole lens triplet shown in Fig. 19.

4. TESTING THE SOFTWARE

4.1 Introduction

A crucial task in the development of any software tool is the validation of the results generated. There are several facets to this task, the most important one being to convince the user as well as the developer that, in cases that the program claims to handle, sensible answers can be obtained from known situations. Another aim of this testing should be to ascertain what

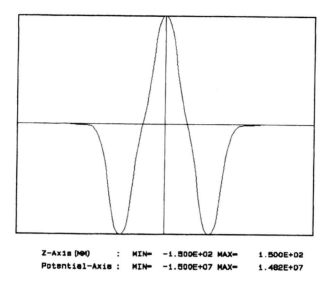

```
Z-Axis (MM)       :  MIN=  -1.500E+02 MAX=   1.500E+02
Potential-Axis :  MIN=  -1.500E+07 MAX=   1.482E+07
```

Figure 21 Example of a plot from program MFPLOT of an axial quadrupole field function from an electrostatic quadrupole lens.

problems the programs can or cannot competently handle, and in what situations are they likely to break down and to offer hints or remedies for any possible failings.

A series of tests on the software packages described in the previous section has therefore been carried out. The bulk of the testing was done on the potential, field and trajectory computation programs, though the other, ancillary programs have all been thoroughly checked and used in many "real" design situations.

Various methods for testing the programs were employed, including: comparing the electrostatic potentials with those computed using analytic models; comparing magnetic axial fields with those generated by other, established numerical methods; comparing the axial fields due to polepiece ellipticity with those computed by a perturbation method; and computing aberrations by direct ray-tracing and comparing those to values predicted using an aberration theory.

Most of the tests were run on an IBM compatible personal computer with a 16 MHz Intel 80386 central processor and 80387 math co-processor. The computer had 4 Mbytes of random access memory and an 80 Mbyte hard disk drive. The programs were configured so that meshes of up to 200,000 nodes could be handled in single precision and up to 100,000 nodes in double precision. A complete analysis generally took between 30 minutes and 2 hours of computer time. The tests described below were all carried out using the double precision versions of the programs.

4.2 Spherical Capacitor Tests

To test the accuracy of the electrostatic formulation including dielectric materials, the following tests were performed. Data were set up for a spherical capacitor consisting of a solid conducting sphere held at potential V_0 and placed concentrically inside a hollow grounded sphere. The potential distribution in this system was computed, firstly with an air gap between the spheres and then with a mixed air/dielectric filling between the spheres.

4.2.1 Spherical Air Capacitor

Initially, the gap between the two spheres was filled with air. The radius of the inner sphere was 5 mm and its potential was 10 V. The inner radius of the hollow, grounded sphere was 20 mm. The potential distribution was computed using double precision variables on three grids with increasing mesh node densities – $21 \times 21 \times 21$ nodes, $41 \times 41 \times 41$ nodes, and $81 \times 81 \times 81$ nodes. (In the latter case, the potentials were computed for an octant of the system on a $41 \times 41 \times 41$ node grid using symmetry conditions on three planes.) The maximum error in the potentials were 0.81%, 0.25%, and 0.07% for the 21^3, 41^3, and 81^3 grids, respectively. [The percentage error, ϵ, in the computed potentials, V_{comp}, compared with the theoretical potentials, V_{theory}, was calculated as:

$$\epsilon = \frac{V_{theory} - V_{comp}}{V_{max} - V_{min}} \times 100 \tag{61}$$

where V_{max} is the maximum potential in the system (10 V in this case) and V_{min} is the minimum potential in the system (0 V in this case).]

4.2.2 Spherical Capacitor with Air/Dielectric Filling

Secondly, the space between the inner and outer sphere was partially filled with a dielectric material with a relative permittivity of ϵ_r. Fig. 22 shows the set-up and dimensions.

It can be shown by applying Gauss's theorem that, at a radius, r, from the center, the voltage, V, is given by:

$$V = V_0 + \frac{V_0 abc}{\epsilon_r(b-c) + b(c-a)} \left[\frac{1}{r} - \frac{1}{a} \right], \quad \text{for } a \leq r \leq b \tag{62}$$

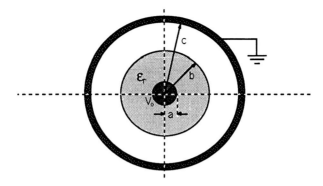

Figure 22 Geometry of spherical capacitor with mixed dielectric insert used to establish the accuracy of the electrostatic potential computation.

and

$$V = \frac{V_0 \epsilon_r abc}{\epsilon_r(b-c) + b(c-a)} \left[\frac{1}{r} - \frac{1}{b} \right], \quad \text{for } b \leq r \leq c \tag{63}$$

The potential distribution was computed on a 41 × 41 × 41 node 3D grid using double precision variables with the following parameters: $V_0 = 10$ V, $a = 5$ mm, $b = 12$ mm, $c = 20$ mm, and $\epsilon_r = 3$. The problem was then solved on a 21 × 21 × 21 node mesh and also an 81 × 81 × 81 node mesh. (Symmetry planes were employed to enable the latter case to be analyzed.) Table 5 summarizes the resulting potentials for each of the meshes at 11 nodes along a mesh line from the center of the spheres to the outer shell. Also shown are the theoretical values at these nodes. Table 6 lists the errors in the computed potentials (calculated using Eq. (61)) at the nodes given in Table 5.

4.2.3 Summary

In all cases, the new programs computed the potentials to an accuracy of better than 2%.

As may be expected, increasing the number of nodes improves the accuracy of the solution. In the system containing electrodes only, the error decreased by a factor of about 3.5 on each successive halving of the mesh spacing. (This is broadly in line with a truncation error of h^2.)

In the system containing the dielectric material, the average absolute error decreased by about a factor of 2.5 on each successive halving of the mesh spacing along each axis. The worst case error fell from 1.7% to 0.3% for a factor of four increase in the number of nodes on each axis (i.e., a 64

Table 5 Computed potentials for the spherical capacitor test with air/dielectric filling on grids with various node densities; analytic values are given in the final column

Rad. (mm)	21 × 21 × 21 grid	41 × 41 × 41 grid	81 × 81 × 81 grid	Exact
20.0	0.00228	0.00000	0.00000	0.00000
18.0	0.40985	0.41239	0.41474	0.41618
16.0	1.38040	1.39896	1.40820	1.41358
14.0	2.61654	2.66207	2.68367	2.69595
12.0	4.23178	4.32636	4.37386	4.40578
10.0	5.14199	5.17567	5.19504	5.20769
8.0	6.39223	6.39365	6.39897	6.40457
6.0	8.38607	8.40294	8.39927	8.39937
4.0	10.0000	10.0000	10.0000	10.0000
2.0	10.0000	10.0000	10.0000	10.0000
0.0	10.0000	10.0000	10.0000	10.0000

Table 6 Percentage error in the computed potentials for the spherical capacitor test with air/dielectric filling on grids with various node densities

Radius (mm)	21 × 21 × 21 grid	41 × 41 × 41 grid	81 × 81 × 81 grid
20.0	−0.0228	0.0000	0.0000
18.0	0.0633	0.0379	0.0144
16.0	0.3318	0.1462	0.0538
14.0	0.7941	0.3388	0.1228
12.0	1.7400	0.7942	0.3192
10.0	0.6570	0.3202	0.1265
8.0	0.1234	0.1092	0.0560
6.0	0.1330	−0.0357	0.0010
4.0	0.0000	0.0000	0.0000
2.0	0.0000	0.0000	0.0000
0.0	0.0000	0.0000	0.0000

fold increase in the total number of nodes). The accuracy of the computation is therefore reduced when dielectrics are introduced into the system. This is due to the approximations made in the new theory regarding the second derivatives of the potentials at the dielectric interfaces. (In fact, Table 6 shows that the worst error in the potentials occurs at the dielectric interface – i.e., a radius of 12 mm – in each of the three grids.) It can be seen, however, that despite these approximations, the new software computes the potentials to a reasonable accuracy.

Although the potential distribution in this test is one-dimensional in a spherical coordinate system, the program computes the solution in a 3D

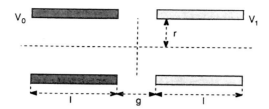

Figure 23 Geometry of the bi-cylinder lens used for spherical aberration computation to test the field computation and ray-trace accuracy.

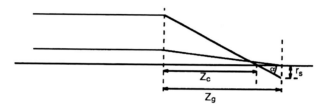

Figure 24 Diagram showing notation used for paraxial rays and focal properties.

Cartesian coordinate system. In this respect the result is still a valid 3D test of the programs' ability to compute and solve the FD equations where the surfaces of the electrodes and dielectrics intersect the grid lines at arbitrary angles.

In conclusion, using a mesh size of $41 \times 41 \times 41 = 68,921$ nodes and utilizing double precision arithmetic, it was possible to compute the electrostatic potential distribution in a system of mixed electrodes and dielectrics to an accuracy of better than 1%.

4.3 Spherical Aberration of a Bi-Potential Lens

As an overall test of the competence of the new 3D electrostatic potential computation and ray-trace programs, the following test was performed. A bi-potential lens, consisting of two hollow cylinders of radius r, length l, separated by a gap g, and with potentials V_0 and V_1, was analyzed (Fig. 23).

An electron traveling from the left of the gap parallel to, and very close to the axis will cross the axis on the right hand side at a position Z_g from the gap, where Z_g is the Gaussian image plane of the lens. If the lens suffers from spherical aberration, electrons entering parallel to the axis, but at a greater radius will cross the axis nearer to the lens on the image side, at Z_c, and will make an angle α with the axis (Fig. 24).

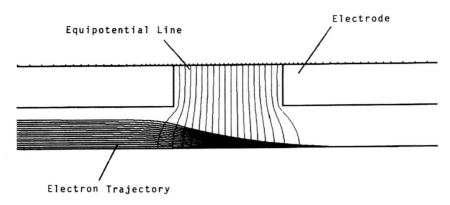

Figure 25 Computed equipotentials and trajectories of electrons being focused in the bi-cylinder test lens.

The off-axis distance, r_s, of such electrons at the Gaussian plane, Z_g can be expressed as a power series in odd powers of $\tan\alpha$ (even powers are excluded due to the rotational symmetry), and is given to the third power of $\tan\alpha$ by:

$$r_s = C_s \tan^3 \alpha + O(\tan^5 \alpha) \tag{64}$$

where C_s is the third order spherical aberration coefficient. By geometry from Fig. 24:

$$\frac{r_s}{Z_g - Z_c} = \tan\alpha \tag{65}$$

therefore

$$Z_c = Z_g - C_s \tan^2 \alpha \tag{66}$$

A plot of the ray's cross-over position, Z_c, versus the square of the tangent of the ray's angle at this position, $\tan^2\alpha$, will yield a straight line with an intercept at $\tan^2\alpha = 0$ of Z_g and slope C_s.

Fig. 25 shows the lens structure used and the computed equipotentials and electron trajectories. Fig. 26 is a close–up view of the trajectories near the focus. The parameters of the lens referred to Fig. 23 are: $g = 50$ mm, $r = 20$ mm, $l = 100$ mm, $V_0 = 1000$ V, and $V_1 = 12454.32$ V.

Fig. 27 is a plot of $\tan^2\alpha$ versus Z_c for trajectories with initial radii up to the inner edge of the cylinders. For small angles the graph is approximately linear, since the third order aberration is dominant in this region. As the

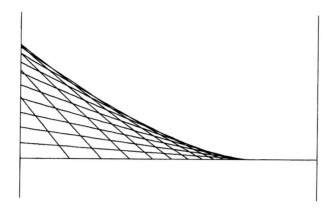

Figure 26 Close-up view of the trajectories near the focus in the bi-cylinder test lens showing the caustic of rays as electrons with wider initial radii cross the axis nearer the object.

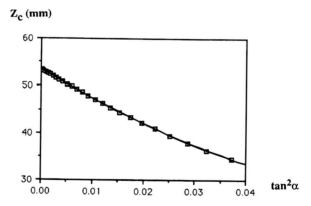

Figure 27 Plot of the square of the tangent of the angle of a ray crossing the axis versus the axial ordinate at which it crosses for the bi-cylinder test lens. The graph was used to deduce the spherical aberration coefficient of the lens.

initial off-axis distance becomes larger (and therefore also $\tan \alpha$), higher order aberrations become significant and the graph deviates from linearity.

From the linear region of the graph at small $\tan \alpha$ the following values for the optical properties were obtained:

$$Z_g = 53.41 \text{ mm}$$
$$C_s = 640.01 \text{ mm}$$

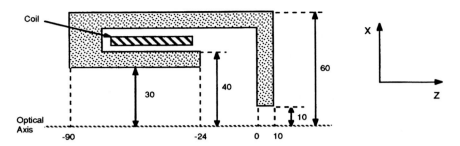

Figure 28 Geometry of the test magnetic lens for which the axial field was computed using the 3D programs ELEC3D and the 2D finite element program MLF2. (Dimensions are in millimeters.)

Using a 2D potential calculation program, together with a paraxial ray-trace and a third order aberration integral (Munro, 1975, 1988), the third order spherical aberration coefficient computed for a Gaussian focal plane at $z = 53.41$ mm was $C_s = 651.2$ mm. Thus, the 3D and paraxial programs are in agreement to within about 2%.

In general, it is far easier to compute the third order optical properties with an aberration integral, but this test has shown that the 3D field computation and ray-tracing programs give reasonable results.

4.4 Axial Field Computation and Ray-Trace in a Magnetic Lens

As a check on the accuracy of the new magnetic formulation, the axial flux density in a round magnetic lens computed with the 3D program was compared with that computed using a 2D finite element program for rotationally symmetric magnetic lenses. This program, *MLF2*, is part of the Imperial College Electron Beam Lithography Package (Chu & Munro, 1981; Imperial College electron optics software catalogue, 1990) and computes the magnetic vector potential, A_θ, in the lens using first order finite elements. This is a well established method and its accuracy has been verified independently (Munro, 1971). It is, furthermore, using an entirely different method to the one the 3D program uses to compute the axial field. The geometry of the test lens is shown in Fig. 28.

The magnetic potentials were computed by each program, using similar numbers of points in the radial and axial directions (40×40). The axial flux density along the axis of the lens was extracted from each analysis and the results are shown in Fig. 29. As can be seen, the two lines are almost coincident.

elliptical defect was introduced in the 20 mm radius bore of the first plate (i.e., an ellipticity of 0.05).

It can be seen from the ray-traces in Fig. 33 that the rays in the two meridional planes focus at different positions. The measured distance, d, between the foci was $d = 55.72$ mm. It can be shown that, for electrostatic lenses, the distance, d, between the line foci in the meridional planes is equal to twice the astigmatism coefficient, A, of the lens. In this case, the astigmatism coefficient, A_1, resulting from a 1 mm elliptical defect in the first bore is computed by the 3D ray-trace to be:

$$A_1 = 27.86 \text{ mm}$$

For an ellipticity of 0.05 in the second bore, the 3D ray-trace predicts an astigmatism coefficient, A_2, of

$$A_2 = 2.93 \text{ mm}$$

The 2D perturbation programs (Munro, 1988) were then used to analyze the same lens and these programs produced the following predictions for the astigmatism coefficients due to a 1 mm elliptical defect in the first and second bore respectively:

$$A_1 = 27.91 \text{ mm}$$
$$A_2 = 2.91 \text{ mm}$$

The values of A_1 differ by 0.1% and the values for A_2 differ by about 0.5%, thereby giving credence to both the 2D perturbation technique and the 3D electrostatic potential and ray-trace programs.

It should be pointed out that the mesh spacing in the lens bore was of the order of 2 mm. A defect of 1 mm is of similar magnitude to this spacing and would, therefore, notably change the FD equations in the bore region from those in the round lens case. If an elliptical defect of 1 μm was introduced into the lens bore in the 3D analysis, the bore radius would be almost constant with angle, θ (see Fig. 32), if viewed on a mesh with a spacing of 2 mm. Any resulting asymmetries in the potentials would be subject to much more error than those quoted for a 1 mm elliptical defect in the bore on a 2 mm spaced mesh. Although the mesh layout can be specified with a certain degree of independence from the structure being analyzed, it should always be defined so that important features of the structure are of similar magnitude to the spacing of the mesh around them.

4.5.2 Magnetic Lens with Elliptical Defect in the Bore

To obtain the astigmatism coefficient in the 2D perturbation model, the required quadrupole fields are computed by assuming that the deviation from roundness is small compared to the bore radius. In this investigation it is shown that, if this condition is satisfied, then the perturbation theory agrees well with a full 3D analysis. It is also seen that if the ellipticity is large, then the results of the 2D and 3D analyses differ. The 3D analysis therefore provides an estimation of the size of the elliptical defect that the perturbation analysis can reliably handle.

For the magnetic analysis, it is less easy to deduce the astigmatism co-efficient from the direct ray-trace, as was done in the electrostatic example above, due to the image rotation in the magnetic field. Instead, the 3D program was used to compute the quadrupole field resulting from an el-lipticity in the bore of a magnetic lens. This was then used to compute the corresponding astigmatism introduced into the lens via an aberration integral. For the 2D calculation, the quadrupole field was computed by the perturbation method and then the astigmatism coefficient was computed with the same aberration integral as in the 3D analysis.

The same lens geometry that was used in Section 4.4 was used again for this test (Fig. 28).

The 2D perturbation program was used to compute the astigmatism that would be introduced by a 1 μm elliptical defect in the bore of the lens. The astigmatism for larger defects was assumed to scale linearly with this figure. The 3D program was used to predict the astigmatism caused by elliptical defects ranging from 0.1 mm to 10 mm. Fig. 34 shows graphi-cally the astigmatism the 2D perturbation analysis and the fully 3D analysis predict for a given ellipticity in the bore of the magnetic lens.

For an ellipticity below 0.1 (i.e., an elliptical defect of 2 mm in a bore of 20 mm), the two methods agree very well. As the ellipticity becomes larger, the results differ – the 3D program predicting a lower astigmatism for a given ellipticity. (The plot of the 2D results is just a linear scaling of the defect computed for an elliptical defect of 1 μm.)

In conclusion, these results once again demonstrate that, for ellipticities below 0.1, the 2D perturbation and 3D analyses are in good agreement. An elliptical defect of 1 mm in a 20 mm bore is extremely large in terms of the roundness which may be achieved with modern machining methods and, therefore, this test seems to indicate that the perturbation programs can be relied upon to assign suitable tolerances for machining errors.

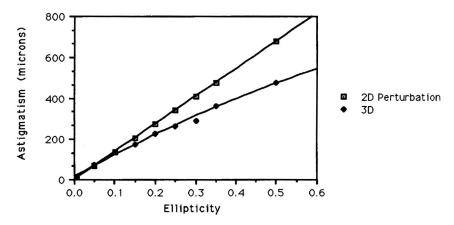

Figure 34 Graph of the astigmatism versus the ellipticity in the bore of the magnetic test lens as computed by a 2D perturbation method and a fully 3D method.

5. ANALYSIS OF PHOTOMULTIPLIER TUBES

5.1 Introduction

Photomultiplier tubes (PMTs) are used to obtain low noise amplification of electromagnetic radiation, especially light. Their general principle of operation is to convert incident photons into electrons by means of a photocathode, P (Fig. 35). The emitted electrons are focused onto a surface, D_1 – the first dynode, which is coated with a material with a high secondary emission coefficient, and emits several secondary electrons for each photoelectron incident on it. The electrons emitted from D_1 are then focused onto D_2, a second dynode, which once more increases the gain of the signal. There are typically 6–10 of these dynodes in a "stack" and they are maintained at progressively increasing potentials (60–100 V per step, say), giving a gain of 10^5 to 10^6. The final dynode focuses the electron signal onto an anode, A, which gives an electrical current signal at the output.

A photomultiplier tube is an ideal detector for converting light into an electronic signal in many applications, because its amplification can range from 10^3 to 10^8 and can, therefore, be matched to most electronic circuits without the need for additional signal amplification. Furthermore, the response time and temporal resolution of the tubes can be of the order of nanoseconds.

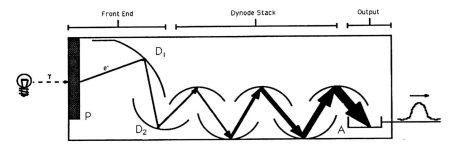

Figure 35 Schematic of the operation of a PMT showing photocathode (P), dynodes (D), and anode (A).

Two major goals in the design of PMTs are: to maximize the collection efficiency between the individual stages, thereby optimizing the sensitivity of the tube; and to minimize the spread in transit times of electrons between individual electron emission and collection surfaces in the tube, thereby optimizing the time resolution. The numerical modeling of the electron optics of the PMT can conveniently be broken down into two stages: the front end, consisting of the photocathode – first dynode region; and the dynode multiplier stack. The modeling of each of these regions will be considered in the following subsections.

5.2 The Front End

The sensitivity of the photomultiplier tube depends critically on collecting sufficient photoelectrons from the photocathode. For many tubes, the photocathode is rotationally symmetric (i.e., cylindrical or spherical) as is the entrance to the dynode stack which is situated co-axially with the photocathode. However, in some tubes the entrance to the first dynode is asymmetric or rectangular and, to minimize the overall diameter of the tube, the stack is sometimes situated off the axis of the front end of the tube (i.e., the entrance to the stack is not co-axial with the photocathode). In these cases a full 3D field and ray-trace computation of the front end region is required to ensure that electrons emitted from all points on the photocathode will reach the first dynode.

Fig. 36 illustrates the model used for such a front end, in three orthogonal views. The chamber containing the photocathode is a grounded cylinder with an off-axis hole drilled through the end face. Through this aperture protrudes an obliquely truncated cylinder at 150 V, joined to a second chamber also at 150 V. The entrance to the second chamber con-

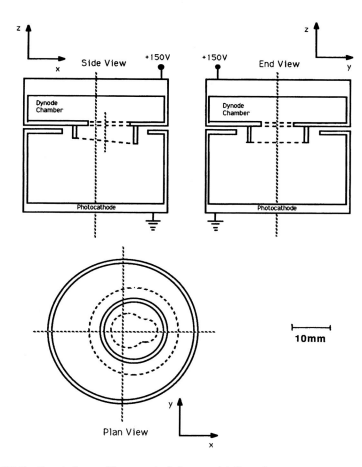

Figure 36 Sectional views of front end of photomultiplier tube.

sists of two overlapping, off-axis holes drilled through its base. The first dynode (not shown) is situated in this second chamber and it is assumed that all electrons passing into this chamber will pass down the full stack. Fig. 37A–C shows the electrostatic equipotential contours in three orthogonal sections through the tube, and trajectories of 1 eV electrons emitted from different points on the photocathode. Fig. 37D shows an orthographic view of the structure and trajectories.[2]

---

[2] This picture – and all subsequent orthographic views – were produced using a commercial 3D solid modeling software package "ModelMate Plus", sold by Control Automation Inc., Florida, USA.

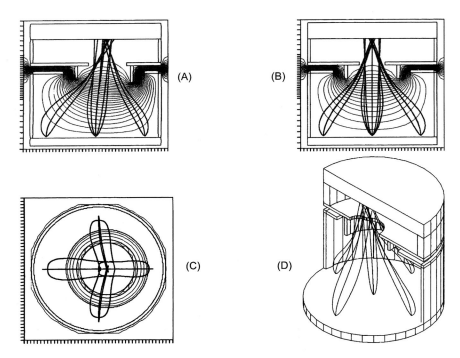

Figure 37 Photomultiplier tube front end with trajectories of 1 eV electrons emitted from the photocathode. (A, B, C) Three orthogonal sections showing equipotential contours; (D) Cut-away orthographic view – drawn using ModelMate Plus.

As can be seen, all the electrons pass through into the second chamber, and hence the collection efficiency of the front end of the tube is good.

The electrons were emitted in bunches of four from five different points on the photocathode [the center $(0, 0)$, $(0, \pm15$ mm$)$, $(\pm15$ mm$, 0)$ in (x, y) – see Fig. 36]. The four electrons in each bunch had a polar angle, θ_p, of 60° and azimuth angles, θ_a, of 0°, 90°, 180°, and 270° (see Fig. 38). Table 7 gives the final positions (at a flat plane 10 mm beyond the entrance to the dynode chamber) and the transit times (in nanoseconds) of the electrons.

From Table 7, one can deduce that all electrons are focused into a region of maximum radius 5.5 mm, the average transit time of all the electrons is 12.9 ns and the maximum transit time difference (i.e., the time difference between the fastest and slowest electron) is 9.8 ns. In a practical tube, the time resolution is limited by the spread in transit time of electrons in the front end region.

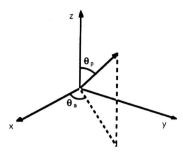

Figure 38 Definition of polar angle (θ_p) and azimuth angle (θ_a) for defining the direction of emission of electrons.

Table 7 Initial positions, (x_i, y_i), final positions, (x_f, y_f), and transit times, (τ), of electrons emitted with 1.0 eV initial energy from the photocathode for various initial azimuth angles (θ_a) (all coordinates are in mm and all times in ns)

	$\theta_a = 0°$		$\theta_a = 90°$		$\theta_a = 180°$		$\theta_a = 270°$	
(x_i, y_i)	(x_f, y_f)	τ	(x_f, y_f)	τ	(x_f, y_f)	τ	(x_f, y_f)	τ
0, 0	4.6, 0.0	8.8	2.1, 2.6	8.8	−0.4, 0.0	8.9	2.1, −2.6	8.8
0, 15	4.8, 0.1	13.4	2.9, −0.9	15.6	0.6, −0.4	13.9	2.5, −0.4	12.4
15, 0	3.0, 0.0	13.1	3.2, 2.3	11.9	2.5, 0.0	11.1	3.2, −2.3	11.9
0, −15	4.8, −0.1	13.4	2.5, 0.4	12.4	0.6, 0.4	13.9	2.9, 0.9	15.6
−15, 0	3.8, 0.0	13.8	4.1, 2.1	15.7	5.5, 0.0	18.6	4.1, −2.1	15.7

In conclusion, this front end has an asymmetric and off-axially situated entrance to the dynode stack to facilitate a more compact, smaller diameter tube construction. Nevertheless, the collection of electrons from the photocathode is good and the time resolution of the front end is better than 10 ns.

5.3 Dynode Stack

Each dynode is made by bending a flat strip of metal round a former and putting metal sides on it to form a bucket shape (Fig. 39). A good approximation to the electron optical behavior of such structures can be obtained by assuming planar symmetry at the center of the bucket and using a 2D numerical field computation and ray-trace. The finite width of the bucket means that this approximation might become less reliable if the bucket is deep compared to its width, or if there is significant electron emission near the edges of the bucket. In these cases, a fully 3D analysis is required to establish how both the collection and the transit times vary for electrons emitted at different positions across the width of the bucket.

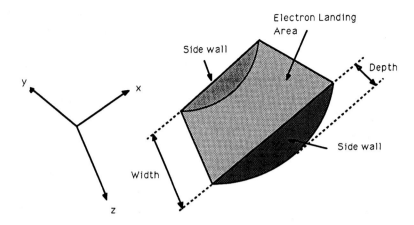

Figure 39 A typical dynode "bucket", with axes used for analysis.

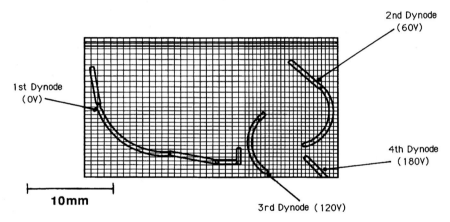

Figure 40 x–y slice through part of the dynode stack showing the structure of the first three buckets and the finite difference mesh used in the analyses.

Fig. 40 is a 2D section (in the x–y plane) through part of a dynode stack, showing the first, second, third, and part of the fourth bucket. The finite difference mesh used to compute the potentials and trajectories is superposed on top of the structure. The width of the stack is 20 mm (± 10 mm in z). A comparison was made of the collection and timing spread of electrons starting at the center of the first dynode ($z = 0$ mm) with electrons starting near the edge of the first dynode ($z = 9$ mm).

Seventeen electrons with initial energies of 1 eV were emitted normal to the surface of the first dynode and in the center of the stack ($z = 0$ mm).

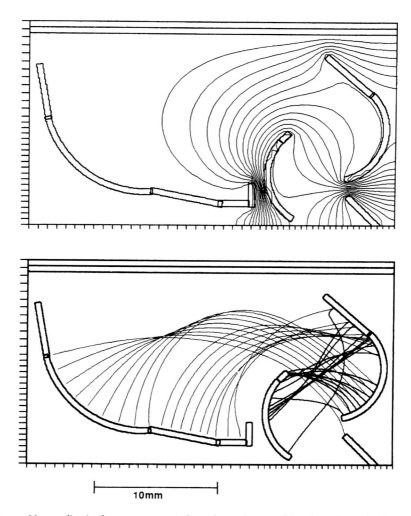

Figure 41 x–y slice in the center, $z = 0$ plane through part of the dynode stack, showing the structure of the first three buckets with: equipotential contours (top); trajectories of electrons from the first to second and second to third dynodes (bottom).

At the point where each of these electrons landed on the second dynode, another electron was emitted normal to the surface, and these electrons were then traced on toward the third dynode. Fig. 41 shows the equipotentials and trajectories in this plane. Table 8 gives the transit times for the electrons traveling from the first to second and from the second to third dynode. From this table, one can deduce that 94% of the electrons emitted

Table 8 Transit times of electrons emitted from the center of the stack with 1.0 eV initial energy. Timings are given in nanoseconds for the trips: from the first dynode to second dynode; from the second dynode to third dynode; and from the first dynode to the third dynode. Trajectories are numbered from left to right on the first dynode

Traj. No.	Dyn. 1 – Dyn. 2	Dyn. 2 – Dyn. 3	Total (D1 – D3)
1	missed	–	–
2	11.58	1.99	13.57
3	11.60	2.35	13.95
4	11.66	2.56	14.22
5	11.56	2.76	14.32
6	11.43	3.03	14.46
7	11.13	3.11	14.24
8	10.74	3.51	14.25
9	10.07	3.55	13.61
10	7.57	4.24	11.81
11	6.80	4.25	11.05
12	6.12	4.32	10.44
13	5.53	4.38	9.91
14	5.02	4.41	9.43
15	4.66	4.28	8.94
16	4.51	4.29	8.80
17	4.12	4.49	8.61

from the first dynode reach the third dynode, the transit time spread from first to second dynode is 7.5 ns and the transit time spread from second to third dynode is 5.5 ns. Furthermore, the transit time spread from first to third dynode is 5.9 ns, so evidently, some compensation of the transit times occurs over the two trips between the three dynodes.

The potentials and trajectories were then computed with the 3D programs but using a 2D model of the stack (i.e., assuming it to have an infinite width in z), and the collection and timing results agreed very well with those for the truly 3D analysis at the center of the stack. Results of the 2D model were verified by comparison with a dedicated piece of software for computing the collection and transit times in PM tube dynode stacks. This software, developed by Thorn EMI Electron Tubes Ltd., Ruislip, UK, assumes 2D planar symmetry and is used as an integral part of Thorn's PM tube design work. Results from the software agree very well with the measured values on tubes constructed from the 2D designs made using their software.

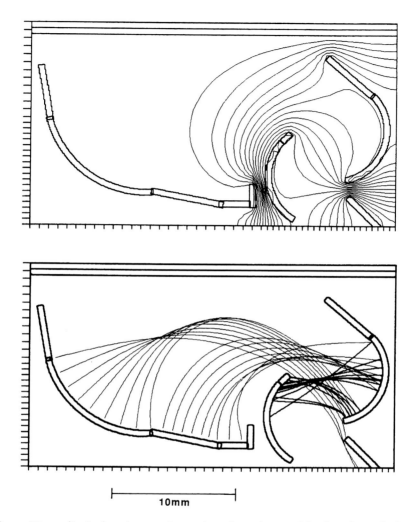

Figure 42 x–y slice in the edge, $z = 9$ mm plane through part of the dynode stack, showing the structure of the first three buckets with: equipotential contours (above); and trajectories of electrons from the first to second and second to third dynodes (below).

Another seventeen electrons of 1 eV initial energy were started in the 3D model at the same x–y coordinates as before, but at $z = 9$ mm from the center of the stack (i.e., 1 mm from the side wall). Fig. 42 shows the equipotentials in this x–y plane. A 2D projection of the trajectories from the first to second and second to third dynode is also shown.

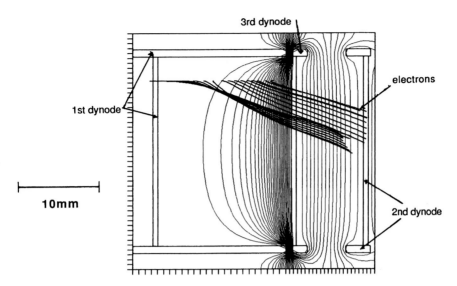

Figure 43 *x–z* slice through part of the dynode stack, showing a plan view of the structure of the first three buckets, equipotential contours and trajectories of electrons being focused toward the center of the bucket as they travel from the first to the second dynode.

Again, electrons were set off normal to the surface of the second dynode, at a point where an electron from the first dynode was incident. In this case, however, the z-position of the electrons is not constant as it was in the case where the electrons were emitted at $z = 0$. (In that instance, the electrons stayed in the $z = 0$ plane because there is no transverse field at the center, due to the symmetry in z.) Fig. 43 is a plan view of the stack, which shows a focussing effect of electrons traveling from the first to the second dynode, forcing them toward the center, $z = 0$ plane of the stack. (The trajectories shown are 2D projections of 3D lines, but the structure and equipotentials are for a 2D slice through the system. This is why some of the electrons appear to stop (and start) in mid air – they are, in fact, incident on a portion of the electrode in a plane which is above or below the chosen viewing plane.) The focussing action demonstrated here can have good and bad effects. On the positive side, electrons are naturally drawn to the center of the stack, where a 2D representation of the potentials and trajectories agree well with the 3D model, thereby making a fast, 2D planar analysis of the stack more accurate than may otherwise be the case. Also, the homogeneity of the chemical layers which receive and emit the electrons is not so crucial across the full width of the bucket – a very uniform

Table 9 Transit times of electrons emitted 1 mm from the side wall of the first dynode with 1.0 eV initial energy. Timings are given in nanoseconds for the trips: from the first dynode to second dynode; from the second dynode to third dynode; and from the first dynode to the third dynode. Trajectories are numbered from left to right on the first dynode

Traj. No.	Dyn. 1 – Dyn. 2	Dyn. 2 – Dyn. 3	Total (D1 – D3)
1	missed	–	–
2	missed	–	–
3	missed	–	–
4	13.56	1.99	15.55
5	13.51	1.97	15.48
6	13.34	2.13	15.47
7	13.01	2.29	15.30
8	12.52	2.42	14.94
9	11.76	2.65	14.41
10	8.65	3.39	12.04
11	7.77	3.56	11.33
12	6.90	3.70	10.60
13	6.13	3.72	9.85
14	5.46	3.73	9.19
15	5.07	3.79	8.86
16	4.99	3.93	8.92
17	4.47	4.13	8.60

coating in the central region will probably yield better results than an average uniformity across the whole bucket width. On the negative side, the crowding of the electrons into the central regions may cause space charge limitations further down the stack where the gain is high.

Table 9 gives the transit times between the first and second dynodes and between the second and third dynodes, for electrons emitted 1 mm from the side wall of the first dynode. This table indicates that, near the edge of the bucket, the collection (82%) is less than that at the center and the overall time spread from the first to the third dynode (7 ns) is greater than that at the center. Clearly, near the edge of the bucket, a 2D model cannot be relied upon to give accurate information of the electron collection or transit times.

The situation is not as bad, however, as it may initially appear. Firstly, of the 17 electrons which successfully get from the first to third dynode in both the center and edge cases, the maximum spread in transit time between the center and edge is 1.3 ns. This is much less than the spread in

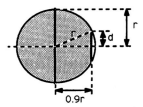

Figure 44 Diagram representing the ratio of the number of electrons along a line at the center of a circular image to the number on a line at a distance 90% of the radius from the center.

times along the length of the first dynode, both for the central and the edge case (5.9 ns and 7 ns, respectively). Secondly, there is a natural tendency for the wide electrons to be focused into the central region of the stack, where the 2D model is valid. (It was found that, for this particular structure, good agreement in the collection and timings could be obtained between the 3D and 2D models for electrons emitted up to 5 mm from the center of the first dynode – i.e., half way to the side wall of the bucket.) Finally, only a small fraction of the electrons are likely to start at such a distance from the center region of the first dynode. This is because the electrons from the photocathode are generally focused into a circular region of the first dynode. By geometry (Fig. 44), the ratio of electrons emitted at the center to those emitted at a distance of 9 mm of a 10 mm radius patch is $d/r = 0.44$. Even this figure is likely to be an overestimate, since the image of the photocathode is usually less than the full width of the first dynode.

In conclusion, the collection from all points on the dynode stack is good and the transit time spread is smaller than that of the front end. A 2D model of the stack is reliable for estimates of collection and transit times provided the electrons from the photocathode are imaged into the central area of the first dynode. The 2D results were found to be unreliable for electrons emitted near the side walls of the bucket.

6. CHARGING EFFECTS ON INSULATING SPECIMENS IN THE SEM

6.1 Introduction

For detailed, quantitative inspection of surface features of specimens using a scanning electron microscope (SEM), it is important to be able to relate the peaks in the collection signals to the edges of the troughs and ridges on the sample in a repeatable way. The science of metrology using the SEM is a field in its own right and much work has been done and is being done to model the imaging processes and correlate them to practical SEM images (e.g., Nunn, 1991, whose help in clarifying and implementing some of the techniques used in this section – especially the Monte Carlo calculations – is gratefully acknowledged). However, in this section, it is demonstrated how the new 3D software could be used to extend the understanding of the relationship between the SEM image and the specimen, when the latter is made of insulating materials.

Generally, secondary electrons are used to provide an image of the surface features, and on a flat, uniform surface, the secondaries are emitted from within a circular region around the location of the primary beam. However, if the primary beam is situated near the edge of a trough in the surface, then it is difficult to predict, *a priori*, where the secondary electrons will be emitted from. This is because some of the primary electrons which undergo elastic collisions may be forward scattered through the side walls and into the bottom of the trough, or backscattered from the bottom of the trough into the side walls (Fig. 45). These forward and back scattered primaries then re-enter the specimen and generate further secondaries from these new sites. In this way, part of the secondary electron signal near the edges of a trough or ridge may well be due to emission from areas far removed from the primary beam position.

Often, it is impractical to coat the surface of an insulating specimen with a layer of conducting material (e.g., gold). In such cases, much of the insulating surface may be exposed to the electron beam, which means that residual charges on the surface cannot be grounded and, therefore, they may build up and generate micro-electric fields around any surface features on the specimen. These micro-fields will, in turn, have some effect on the paths of the secondaries leaving the specimen and hence alter the collection distribution as a function of the surface charge.

In the special case of on-line electron beam inspection of passivated integrated circuits, the primary beam energy used must be low (typically below 1–2 keV), to ensure that no damage is done to the electronic structure of the devices on the chip. The use of a low primary beam energy may result in the secondary electron emission coefficient being greater than unity, causing more secondary electrons to leave the sample than there are primary electrons incident on it. In this case, the specimen will acquire a positive charge, causing some secondary electrons to be attracted back to the surface of the sample and not be detected.

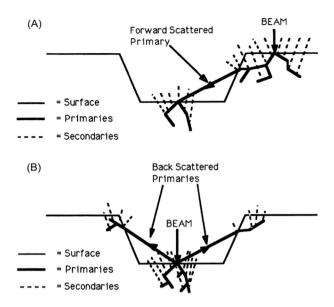

Figure 45 Elastic scattering of primary electrons near a trough. (A) Forward scattering through the side walls. (B) Back scattering from the bottom surface.

In order to gain some insight into the physics of these processes and how they may affect the imaging of insulating specimens in the SEM, we aim to use the new 3D software to simulate a single line sweep across a trough in a charged dielectric specimen.

6.2 Strategy of the Method

Two distinct line scan simulations have been performed: the first was using a high energy primary beam (10 keV); the second was using a low energy primary beam (800 eV). In each analysis, the general methodology was the same, though the nature of the charging and the primary beam penetration into the specimen (and hence the secondary electron emission pattern) will be markedly different. The simulation of the line sweep was broken down into three separate tasks. Essentially one needs to know where the secondaries will emerge from the surface, how they are affected by the external fields when they are emitted, and what contribution they make to the signal. This information must be obtained at many positions on the surface of the sample and the data must be used to produce a simulated line scan. The individual tasks are outlined in the following subsections; a description of the methods used and the approximations made is also given.

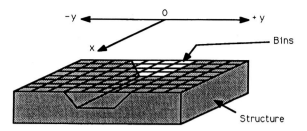

Figure 46 Diagram illustrating the "bins" used for counting the emitted secondary electrons on the surface of the sample.

6.2.1 Primary Electron – Sample Interactions

The first task was to model the interaction of the primary beam with the specimen and the subsequent production and distribution of secondary electrons. This was done using a Monte Carlo technique (Joy, 1988). The software which performed these calculations, IC-SE, was derived from a program, SEGEN (Joy, 1989), written by David Joy in PASCAL, which was converted to FORTRAN 77 and modified slightly. The program subdivided the area of the sample into a series of "bins" which were fixed in space, relative to the surface of the sample, and symmetric about the y-axis (Fig. 46).

The primary beam was placed at a certain y-position (on the $x = 0$ line) and a Monte Carlo calculation (Joy, 1988) traced the primary electrons as they were scattered within the sample and forward and back scattered to other regions near the trough. At every time step on the primary electron path, the energy of the electrons was computed (according to the Bethe assumption of constant energy loss from point of impact at the surface), and this energy was used to compute the number of secondary electrons generated in this time interval. The number of generated secondaries was then multiplied by the probability of their escaping from the depth below the surface at which they were generated; the bin vertically above the point of creation was incremented by this product.

When all the primary electrons had been traced, the total contents of the bins provided a secondary electron emission distribution over the surface of the sample corresponding to a certain primary beam position. Repetition of this step for many primary beam positions yielded a series of distributions as the primary beam was scanned across the sample.

6.2.2 Three-Dimensional Field Computation and Ray-Trace

The next step was to model the electrostatic micro-fields around the surface features on the chip which were created by the charges deposited by the primary beam, and to trace the secondary electrons in these fields after they emerged from the sample surface.

The potentials were computed using the new 3D electrostatic package, ELEC3D, which can handle charged dielectric materials. For the purpose of these calculations, it was assumed that the primary beam had made many scans of the sample at a low current, so that a uniform charge was already deposited on the surface of the dielectric sample. In the case of the high primary beam energy, this charge was assumed to be negative, because the secondary electron emission coefficient of most materials is less than one for this beam energy (i.e., the number of emitted secondary electrons is less than the number of incident primary electrons). Conversely, at the low beam energy, a uniform positive charge was placed on the sample surface because, at this primary beam energy, the secondary electron emission co-efficient was assumed to be greater than one (i.e., more secondary electrons leave the sample than there are primary electrons incident upon it). Assuming this uniform charge distribution, the potential distribution resulting from these charges was computed. No external field, which may have been created at the surface by a secondary electron collector, was considered.

The trajectory computation program, ETRAJ, from the new 3D soft-ware package was customized and used to compute the trajectories of a bunch of electrons from each point on the line scan. The micro-fields around the surface features of the sample created by the charge distribu-tion affect the paths of the emitted secondaries, and some of secondaries will return to the sample. In the absence of a collector field, an electron was assumed to be collected if it didn't return to the sample surface. Those secondaries which were collected were recorded and, when expressed as a percentage of the emitted electrons, gave a figure for the collection effi-ciency at a point on the sample. A value for the collection efficiency was obtained at many points on the line scan to give a collection distribution across the sample.

The emission and subsequent recapture of some of the secondary elec-trons by the sample would, in practice, cause a redistribution of the surface charges. It was assumed, however, that the current in the primary beam of the single line scan that was simulated was low and, therefore, the redistri-bution of the charges was considered to be small enough (in comparison with the charges already present on the surface) to be neglected.

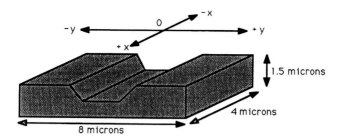

Figure 47 Perspective view of the sample block over which the simulated line scan was made.

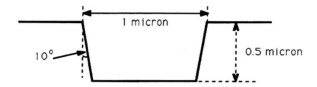

Figure 48 Cross section of the trough in the sample block, showing the dimensions used in the simulation.

6.2.3 Simulated Line Scan

The final task involved the convolution of the secondary electron emission distribution at each step on the scan with the collection efficiency distribution, in order to obtain a simulated line scan. A program, CONV, was written which performed this computation by taking the emission distribution for a primary beam position, y' say, on the scan line and integrating the product of the collection efficiency at any point, y, with the secondary electron emission at y, for all y values over the scan. This gave a signal value at y' and when this was repeated for all $y's$ on the scan, a simulated signal was obtained corresponding to the specimen being scanned.

6.3 Description of the Insulating Sample

A slab of silicon, 4 μm × 8 μm in x–y, and 1.5 μm high in z was taken as a sample. In this block, a 4 μm long, 1 μm wide and 0.5 μm deep trough, centered on the x-axis, was cut (see Fig. 47), with a cross section as shown in Fig. 48.

The block was placed on a grounded plate in a box 8 μm × 8 μm × 5 μm (x, y, z) upon which Neumann boundary conditions were assumed.

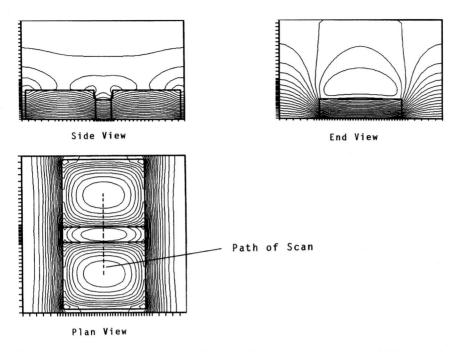

Figure 49 Equipotentials in three orthogonal planes through the center of the trough for a negatively charged dielectric sample.

6.4 High Incident Beam Energy Line Scan

Firstly, we consider the computations which were performed to simulate the signal obtained when a single sweep of a 10 keV primary beam is made across the surface of a negatively charged dielectric sample. In this section, the steps outlined in Section 6.2 are described in detail, for the high energy primary beam case.

6.4.1 3D Potential Computation

A uniform negative, surface charge density of 10^{-11} C/μm^2 was placed on the sample (roughly corresponding to scanning a 100 μm $\times$ 100 μm area of the sample at a video frequency with a 50 pA beam current for half an hour), and the electrostatic potential distribution was computed in each case, using the new 3D software, on a $40 \times 40 \times 40$ node grid.

Equipotentials in three orthogonal planes through the center of the trough were obtained using program ECONT and are shown in Fig. 49.

6.4.2 Secondary Electron Emission Computation

A primary beam of energy 10 keV and diameter 0.01 μm was placed at a point on the silicon target (atomic number 14, atomic weight 28, density 2.4 g/cc). The program IC–SE was then used to compute the secondary electron emission distribution using a bunch of 5000 primary electrons for the Monte Carlo computation. The emission distributions were obtained for many points on the scan line (over the range -1.5 μm $\leq y \leq 1.5$ μm). Fig. 50 shows log plots of secondary electron emission over the scan line for three different primary beam positions on that line.

It can be seen that when the primary beam is placed away from the upper edge of the trough (Fig. 50A), the major secondary electron emission is from around the primary beam. There is, however, significant emission from the edge of the trough nearest to the primary beam and also a noticeable signal from the edge furthest from the beam. When the beam is placed on the trough edge (Fig. 50B), the secondary emission is peaked at the incident beam site, but significant emission occurs from the bottom of the trough near the edge where the primary beam is placed. The emission rapidly falls off across the trough, but there is again a visible signal from the opposite edge of the trough. These emission characteristics are presumably due to the forward scattering of primary electrons through the side wall and into the trough (as depicted in Fig. 45A). When the beam is placed on the bottom surface of the trough (Fig. 50C), the emission is essentially localized around the primary beam site. Some emission from either side wall can be discerned, due to primaries backscattering from the bottom surface into the side walls, as depicted in Fig. 45B.

6.4.3 Collection Efficiency Computation

Bunches of secondary electrons with initial energies of 1 eV were set off at many points on the scan line in cones with the cone axis normal to the sample surface (Fig. 51). Each cone contained 36 electrons distributed through 360 degrees, and the cones had polar angles (θ_p in Fig. 51) ranging from 10 to 80 degrees in 10 degree intervals.

The trajectories of these electrons were computed and the number which returned to the surface was counted. The collection efficiency (i.e., the fraction of electrons which did not return to the sample surface) was computed at each point on the scan line in the range -2.0 μm $\leq y \leq 2.0$ μm. Fig. 52 shows a 3D orthographic view of the sample with a bunch of trajectories which have been emitted from the center of the trough.

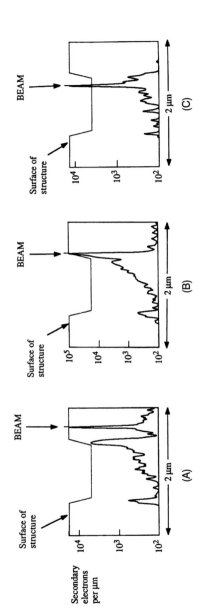

Figure 50 Log plots of secondary electron count across the sample, for three different positions of a 10 keV incident beam: (A) Outside trough; (B) on edge of trough; and (C) on the bottom surface of the trough. The count is larger when the beam is on the edge of the trough due to emission through the side wall.

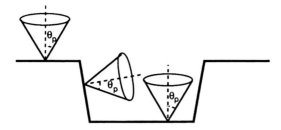

Figure 51 Schematic diagram showing emission cone geometry at points on the top surface, side wall, and bottom surface of the trench.

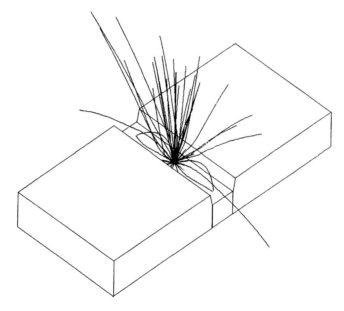

Figure 52 Orthographic view of the negatively charged sample showing the trajectories of a bunch of electrons emitted from the center of the trough showing how some electrons are forcedback to the sample surface.

When the collection efficiency has been computed for all points on the scan line, a collection distribution (Fig. 53) can be plotted.

Fig. 53 shows that all the emitted secondaries are collected as the trough is approached, because there is a repulsive field on the top edge of the trough. There is a dramatic drop in collection (down to 15%) as the trough is entered. Once inside the trough, the collection begins to rise, reaching a peak in the center, but even at the center, the collection is only 50%.

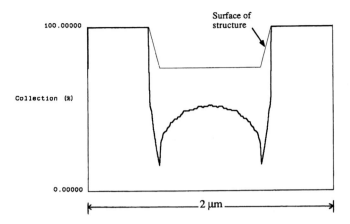

Figure 53 Percentage of electrons collected as a function of position across the complete line scan for a negatively charged sample. The collection is worse near the bottom corners of the trough.

Inside the trough, some of the emitted secondary electrons will hit the side walls. Also, some are trapped in the trough since the repulsive fields (due to the negative charges on the upper edges of the trough) force electrons emitted from sites on the bottom surface back down into the trough. (This latter effect can be discerned from the trajectory plots in Fig. 52.) It can be assumed that the collection would continue to rise until it reached the level on the top surface if the trough width was large compared to its depth.

6.4.4 Simulated Line Scan

The program CONV was then used to convolve each of the emission distributions (as those shown in Fig. 50) with the collection distribution (shown in Fig. 53), to obtain a simulated line scan signal as the primary beam is swept across the trough. The results are shown in Fig. 54.

As can be seen from Fig. 54, the signal rises steadily as the trough is approached, reaching a peak on the upper edge. There is then a dramatic fall in intensity as the trough is entered, followed by a minor recovery toward the center of the trough. However, the signal inside the trough is much lower than that on the upper surface. In this case, therefore, one may expect the edge of the trough to be marked by a bright peak with a dark line just inside the trough, and the center of the trough to be relatively dark.

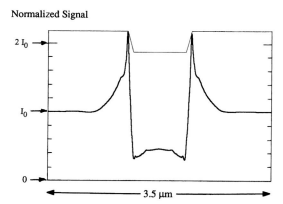

Figure 54 Simulated line scan signal across the sample using a 10 keV primary beam, predicting a high signal near the edge of the trough.

6.5 Low Incident Beam Energy Line Scan

Next, we consider the computations which were performed to simulate the signal which may be obtained when a single sweep of an 800 eV primary beam is made across the surface of a positively charged dielectric sample. In this section, the steps outlined in Section 6.2 are described in detail, for the low energy primary beam case.

6.5.1 3D Potential Computation

For this case, a uniform, positive surface charge density of 10^{-11} C/μm^2 was placed on the sample and the electrostatic potential distribution was again computed using the new 3D software. The equipotentials plots shown in Fig. 49 are of identical form for the positively charged case, but of opposite sign.

6.5.2 Secondary Electron Emission Computation

A primary beam of energy 800 eV and diameter 0.01 μm was placed at a point on the silicon target and the program IC-SE was used to compute the secondary electron emission distribution using a bunch of 5000 primary electrons. Emission distributions were obtained for many points on the scan line (over the range -1.5 μm $\leq y \leq 1.5$ μm). Fig. 55 shows log plots of secondary electron emission over the scan line for three different primary beam positions on that line.

It can be seen that when the primary beam is placed away from the upper edge of the trough (Fig. 55A), the secondary electron emission is

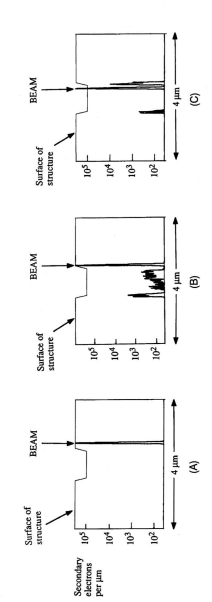

Figure 55 Log plots of secondary electron count across the sample, for three different positions of an 800 eV primary beam: (A) On top surface; (B) on side wall; and (C) on bottom surface of trough. Forward scattering is less evident and backscattering is more prominent at the lower beam energy.

localized around the primary beam. When the beam is placed on the trough edge (Fig. 55B), there is secondary emission from the bottom of the trough and the opposite side wall. This is presumably due to the forward scattering of primary electrons through the side wall and into the trough (as depicted in Fig. 45A). When the beam is placed on the bottom surface of the trough (Fig. 55C), there is significant emission from the primary beam site but also emissions from both side walls. This is presumably due to primaries backscattering from the bottom surface into the side walls, as depicted in Fig. 45B.

It is interesting to contrast the emission distributions at this low beam energy with those at the higher incident energy (Fig. 50). Because the primary electrons penetrate further into the sample at the higher voltage, the forward scattering through the side walls is more pronounced (Figs. 50A and 50B). However, for the same reason, the backscattering from the bottom surface into the side walls (Fig. 50C) is less evident.

6.5.3 Collection Efficiency Computation

Bunches of secondary electrons with initial energies of 1 eV were set off at many points on the scan line in cones with the cone axis normal to the sample surface (Fig. 51). Each cone contained 36 electrons distributed through 360 degrees, and the cones had polar angles (θ_p in Fig. 51) ranging from 10 to 80 degrees in 10 degree intervals.

The trajectories of these electrons were traced using the new 3D ray-trace, and the collection efficiency (i.e., the fraction of electrons which did not return to the sample surface) was computed at each point on the scan line in the range -2.0 μm $\leq \gamma \leq 2.0$ μm. Fig. 56 shows two 2D sections through the sample with 2D projections of the trajectories.

When the collection efficiency has been computed for all points on the scan line, a collection distribution (Fig. 57) can be plotted.

Fig. 57A shows that the collection efficiency is about 90% at a distance of 2 μm from the trough and falls off steadily as the trough is approached, reaching a value of about 68% at the edge. Presumably, this is because electrons are forced back to the surface, due to a large attractive field on the top edge of the trough (resulting from the positive charges on the surface). There is a dramatic drop in the collection as the trough is entered, but a rapid rise once inside, rising to a peak in the center. It can be assumed that the collection would continue to rise until it reached the level on the top surface if the trough width were large compared to its depth.

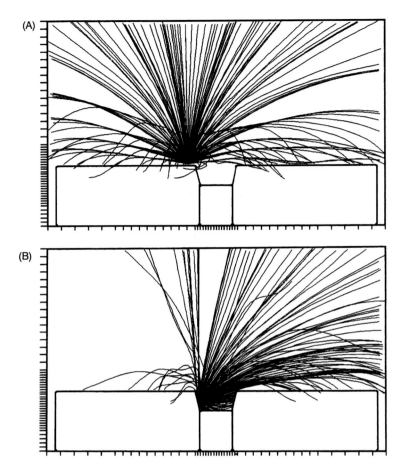

Figure 56 Two dimensional projections of secondary electron trajectories emitted from two points on the positively charged sample surface. (A) Emission point on the upper surface showing electrons returning to the positively charged surface. (B) Emission point on side wall of trough showing electrons curving up out of the trough.

Fig. 57B shows an enlarged view of the collection distribution around the trough. The reason for the pronounced fall in the collection near the top of the side wall and subsequent recovery near the bottom of the side wall seems slightly puzzling. A certain drop in collection is expected from sites on the side walls and near the bottom corners of the trough, since many of the electrons are emitted toward a surface (i.e., downward on the side wall sites, or toward the side wall on the bottom surface sites). If this were the sole explanation for the drop in collection, however, one

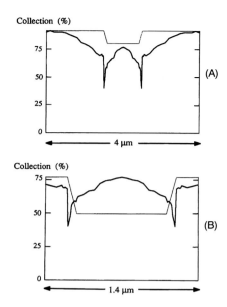

Figure 57 Percentage of electrons collected from the positively charged sample as a function of position. (A) Across the complete line scan. (B) Enlarged view near the trough, showing a steady drop in collection as the trough is approached.

would expect a fall-off down the side wall with a minimum at the bottom (since the emission site is in the corner between two faces), and a rise from there to the center of the trough. This is almost exactly the picture one obtains in the case of the negatively charged sample at higher beam voltage (Fig. 53).

The behavior of the collection distribution in the positively charged case is, therefore, probably due to the attractive fields on the top corners of the trough. Electrons emitted on the side wall "see" two attractive field regions above them – one on either top edge of the trough. Near the top of the wall, many electrons emitted upward immediately feel the effect of the field on the corner directly above them and are attracted back to the surface, while those emitted horizontally are accelerated toward the field on the opposite edge. Some of latter electrons hit the opposite wall and some just miss it, but are traveling parallel to the surface and are subsequently captured further along the top surface. Near the bottom of the wall, many electrons which are emitted upward are accelerated by the field vertically above them, but are then traveling fast enough when they leave the trough to escape from the surface. Some electrons which are emitted

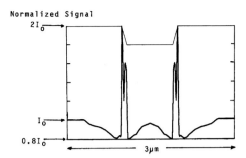

Figure 58 Simulated line scan signal across the positively charged dielectric sample with a 800 eV incident beam, with the edge of the trough marked by a double high peak in signal.

horizontally or downward will hit the bottom surface or the opposite wall, but some (especially those with a low initial energy) will curve up toward the attractive field on the opposite top edge. Those that miss the opposite edge are then traveling fast and at a larger angle to the horizontal, and so are less likely to become trapped on the upper surface than their counterparts emitted further up the wall. (The curving up of downward emitted electrons is evident in Fig. 56B.)

6.5.4 Simulated Line Scan

The program CONV was then used to convolve each of the emission distributions (as those shown in Fig. 55) with the collection distribution (shown in Fig. 57), to obtain a simulated line scan signal as the primary beam is swept across the trough. The results are shown in Fig. 58.

As can be seen from Fig. 58, the signal drops off as the trough is approached and then rises to a peak at the top edge. There is then a dip in the signal (due to the drop in collection from sites near the top of the side wall, as discussed in the previous section) and then a second peak near the bottom of the side wall. The signal then rises toward the center of the trough, where the level is comparable to that on the top surface at some distance from the trough edge. Thus, the model is predicting that the trough edge is marked by a double peak in intensity, with a dark line on either side of the edge, and that the center of the trough should be relatively bright.

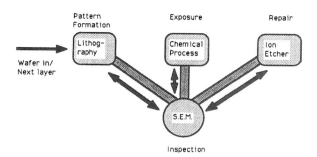

Figure 59 Schematic diagram of an SEM integrated into a wafer fabrication line so that the sample can be inspected at intermediate stages of production.

7. A DESIGN STUDY OF THREE MAGNETIC IMMERSION LENSES

7.1 Introduction

During the manufacture of integrated circuits, it is often necessary to view the wafer at intermediate stages of the processing. One may, for example, imprint a pattern on the chip surface, observe the pattern with a scanning electron microscope (SEM), expose the chip to chemical vapor or implant dopants into it, inspect the chip again to check that the exposure has been successful, correct any defects with an ion implanter, review the pattern, and then continue to the next layer. Traditionally, the inspection stages mean the removal of the wafer from the line to a separate location for viewing in the SEM. This requires the use of extra high grade clean rooms or the use of special transport boxes for the transfer of the wafers between machines. It would be preferable, therefore, to incorporate a microscope into the production line so that the wafers can pass from micro-lithography to chemical processing to ion etching machines via an SEM without leaving the vacuum (Fig. 59).

In a commercial SEM, the objective lens is usually a round magnetic type which is placed above the specimen so that the latter is in a field free region. The specimen chamber is often a rectangular box and is magnetically isolated from the final lens, in the sense that its shape does not perturb the focussing properties of the lens. There are several advantages in having a rectangular chamber. Firstly there is the question of cost: rectangular lengths of metal can be forged in a continuous process and cut to length; the sides of the box can then be welded together to form a chamber. This method dramatically reduces the amount of turning and grinding (both expensive engineering processes) that have to be performed in the chamber construction. Secondly, a rectangular box gives a larger practical working volume for stages, probes, etc. inside the chamber. Thirdly, the flat faces of the box facilitate easier airtight coupling of components to the chamber.

For the in-line inspection system, there are several advantages to be gained by bringing the specimen toward, and even inside the final lens. Firstly, the focal length of the lens is reduced and, therefore, the spherical and chromatic aberration coefficients are correspondingly lower. Secondly, it makes the column more compact, which means a physical saving of space but, more importantly, gives less time for Coulombic interactions between electrons to blur the beam (this is especially important for the low energy beams used for integrated circuit inspection – typically around 1 keV). Thirdly, it makes the column easier to shield from random external or stray electromagnetic fields, since there is not a long throw distance between the lens and specimen, but instead, the specimen is inside a large iron lens.

There are, however, certain drawbacks with this approach. If the final lens is not rotationally symmetric, asymmetric aberrations and displacements are introduced into the beam. For a specimen-in-lens system (in which the final lens is effectively the specimen chamber), there is clearly a conflict between having a rectangular chamber, and not perturbing the beam too much by having a non-rotationally symmetric final lens. The problem is compounded in this particular application because the specimens (8 inch diameter wafers) are so large that they cannot be inserted into the bore of the lens. Instead, the wafers must enter the lens and be placed on the optical axis via side slots cut into the outer casing of the lens. There is also a danger of saturation occurring, since the lens may be required to run at a reasonably high excitation to achieve a short focal length. Other problems with an immersion lens are that the stage and specimen cannot be made of ferrous materials, since their presence in the lens would cause distortions of the focussing field. Furthermore, the problem of collecting the low energy secondary electrons from the sample requires careful thought. This is because the secondary electrons are emitted in a magnetic field and, hence, cannot be "sucked away" to the side by electrostatic collectors, as is often possible in the conventional SEM. There are, however, viable schemes for collecting the low energy secondaries. In one scheme, devised by Plies (1990), the secondaries spiral back through the bore of the magnetic lens, and are collected electrostatically outside the lens in a field-free region. The use of this type of objective lens in a scanning electron microscope, where the polepiece and specimen chamber form an integral unit, has been reported by Shao and Lin (1989). In the present design study of such systems, three immersion lens designs were considered, ranging from an essentially round, slotted lens to a rectangular box with a circular aperture and pole-piece gap region. The designs for these three lenses were kindly proposed by Dr. P.R. Thornton of Hitachi Instruments Inc., San Jose, California. The optical properties of each lens have been computed, including the optical defects introduced by the departure from rotational symmetry. The feasibility of correcting these optical defects with stigmators and alignment coils has been assessed, as has the relative merits of each lens with regard to its dual role as a specimen chamber.

7.2 Description of the Immersion Lens Designs Considered

This section describes the physical structure of the three lenses and shows magnetic scalar equipotentials in selected sections through each. The lenses were named MODEL1, MODEL2, and MODEL3.

7.2.1 Lens MODEL1

This lens (Fig. 60) consists of a round immersion lens of outer diameter 350 mm and overall height 175 mm, with a polepiece gap of 50 mm. The upper polepiece has a bore radius of 10 mm and the lower polepiece is solid. Through the sides of the lens are milled two diametrically opposed, rectangular slots, 10 mm high and wide enough to allow an 8 inch (200 mm) diameter wafer to be passed right through the lens. Two round evacuation ports of diameter 70 mm are also drilled in the base of the lens.

With the coils in place, this lens offers minimal practical space for probes and detectors. Also, the round surfaces mean the coupling of external components to the chamber is more complicated than for flat sided chambers.

Figs. 61A–C are three orthogonal sections through the lens, showing the polepiece structure and magnetic scalar equipotential contours. Fig. 61D is a cut-away 3D solid model of the lens with electrons of 1 keV focussing at the wafer plane. Things to note from the equipotential plots are: the evident flux leakage from the ports and slots; and the high contour density in the gap region, indicating a large field in this area.

7.2.2 Lens MODEL2

This lens (Fig. 62) has a cylindrical region of iron of outer diameter 100 mm surrounding the coil and the upper polepiece bore. This unit is placed on top and to one side of a hollow rectangular box with a cylindrical hole of diameter 85 mm in its top surface. The dimensions of the box

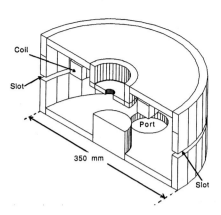

Figure 60 Cut-away 3D view of lens MODEL1 showing the two slots through which the wafer is passed in and out of the lens and the ports for evacuating the lens/chamber.

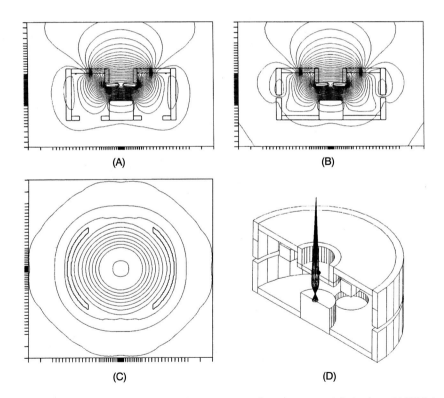

Figure 61 Polepiece structure and magnetic scalar equipotentials in lens MODEL1: (A) Sectional plan view; (B) sectional end view; (C) sectional side view; (D) cut-away 3D model with electron trajectories.

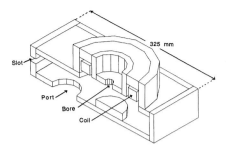

Figure 62 Cut-away 3D view of lens MODEL2 showing the single slot at the left of the lens for wafer entry/exit and the single evacuation port.

are 325 mm (length) × 256 mm (width) × 76 mm (height). Below the bore is a solid cylinder (diameter 38 mm), which is attached to the base of the box and forms a solid lower polepiece. As in MODEL1, the gap is 50 mm and the upper polepiece bore radius is 10 mm. In the base of the lens, to one side of the optical axis, an evacuation port of diameter 70 mm has been drilled, and in one end face there is a rectangular wafer slot, 210 mm wide and 13 mm high.

As a specimen chamber, this lens offers a large working area for the handling of the wafer between the slot and the gap. Space is still limited, however, for the placement of detectors or probes near the polepiece gap.

Figs. 63A–C are three orthogonal sections through the lens, showing the polepiece structure and magnetic scalar equipotential contours. Fig. 63D is a cut-away 3D solid model of the lens with 1 keV electrons being focused at the wafer plane.

7.2.3 Lens MODEL3

This lens (Fig. 64) is essentially a square box with two cylindrical inserts inside it. One insert (diameter 50 mm) contains the upper polepiece of the lens and has the coil wound round it. The second insert is a disk (diameter 38 mm) which acts as the solid lower polepiece and defines the lens gap region. (The gap and upper polepiece bore radius are, once again, 50 mm and 10 mm respectively.) The dimensions of the box are 325 mm (length) × 254 mm (width) × 154 mm (height). As in MODEL2, there is a 70 mm diameter evacuation port in the base of the lens and a rectangular wafer slot 210 mm wide and 13 mm high in one end face.

As a specimen chamber, this lens offers the maximum working volume for handling the wafer and for placing detectors or probes near the pole-

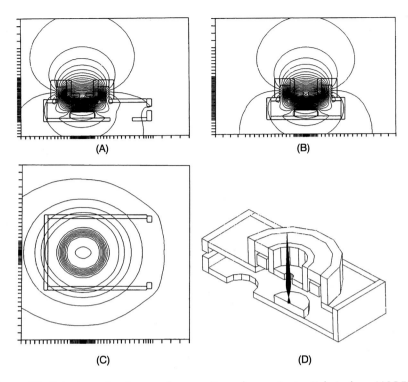

Figure 63 Polepiece structure and magnetic scalar equipotentials in lens MODEL2: (A) Sectional plan view; (B) sectional end view; (C) sectional side view; (D) cut-away 3D model with electron trajectories.

piece gap. Its flat sides also facilitate easy airtight coupling of components to the chamber.

Figs. 65A–C are three orthogonal sections through the lens, showing the polepiece structure and magnetic scalar equipotential contours. Fig. 65D is a cut-away 3D solid model of the lens with 1 keV electrons focussing at the wafer plane.

7.3 Optical Properties of the Lenses

To compute the optical properties, the program MAXIAL (see Section 3.7) was used to extract the axial and near axial magnetic field components from the 3D potential distributions. These data were read directly by program OPTISF (part of the STRAYFIELD Package; Zhu, 1989) which computed the optical properties for specified imaging conditions.

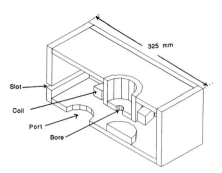

Figure 64 Cut-away 3D view of lens MODEL3 showing its box-like nature and the wafer slot and evacuation port to the left of the optical axis.

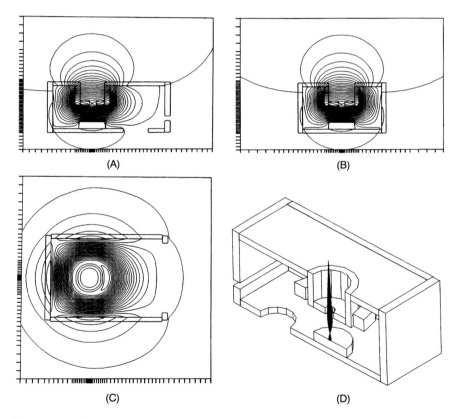

Figure 65 Polepiece structure and magnetic scalar equipotentials in lens MODEL3: (A) Sectional plan view; (B) sectional end view; (C) sectional side view; (D) cut-away 3D model with electron trajectories.

The beam energy was set at 1 keV (to avoid damage to the wafer) and the computed optical properties of the lenses are given in Table 10 for an object – image distance of 150 mm and a beam semi-angle at the image of 3 mrad.

The first order focussing properties for each lens are similar, as are the spherical and chromatic aberration coefficients. The excitation of each lens (on average, about 450 Ampere turns) was low enough for saturation effects to be ignored. For a 3 mrad aperture angle, the diameter of the spherical aberration disk is less than 0.001 μm, and assuming an energy spread of 1 eV, the diameter of the chromatic disk is about 0.06 μm in each case.

In general terms, the departure from rotational symmetry in the pole-piece structure may cause image shift, coma, and astigmatism, and the computed values of these are also given in Table 10.

It can be seen that no image shifting occurs with MODEL1, since the two diametrically opposed slots in the round lens do not produce any deflection field component. The image shift is significant with MODEL2 and larger still with MODEL3. The image shift in MODEL2 ($\Delta x = 55$ μm, $\Delta y = 14$ μm) can easily be corrected using direct current offsets on existing deflectors. The image shift in MODEL3 is a more serious problem ($\Delta x = 168$ μm, $\Delta y = 66$ μm), but with a large scan field deflection system, as may be used in a wafer inspection column, even these shifts should be tolerable, as they are small compared with, for example, a 1 cm × 1 cm field.

The coma in each lens is negligible.

The astigmatism increases through MODEL1–MODEL3, but appears to be correctable by conventional stigmators in each case. With two diametrically opposed slots in a round lens (as is essentially the case for MODEL1), it would be expected that astigmatism would be the dominant asymmetry aberration. This is because the two slots introduce a quadrupole field component which is proportional to the astigmatism. Furthermore, the lens MODEL1 has been analyzed with only the wafer slots and then with only the evacuation ports perturbing the rotational symmetry. Results of these analyses show that the major contributor to the astigmatism of this lens is the wafer slots; the evacuation ports introducing only about a third as much astigmatism.

Figs. 66, 67, and 68 are spot diagrams for the lenses MODEL1, MODEL2, and MODEL3 respectively. For each lens the diagrams show the predicted image of a point source for three different beam energies ($E_0 - \frac{1}{2}\Delta E$, E_0, and $E_0 + \frac{1}{2}\Delta E$, where E_0 is the nominal energy of the beam and ΔE is the energy spread of the beam). Two grids are shown: the

Table 10 Computed optical properties of the immersion lenses MODEL1, MODEL2, and MODEL3

Lens model	Excitation	Mag.	Rot.	Cs	Cc	Image shift		Coma	Astigmatism
	(at)		(deg)	(mm)	(mm)	x (μm)	γ (μm)	(μm)	(μm)
1	522	−0.16	112.8	23.1	17.9	0	0	0.000	0.4
2	415	−0.18	107.8	30.3	20.7	−55	−14	0.000	1.6
3	433	−0.18	110.0	27.7	20.1	−168	−66	0.003	1.4

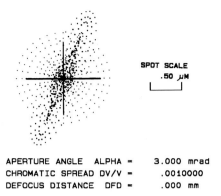

APERTURE ANGLE   ALPHA =   3.000 mrad
CHROMATIC SPREAD DV/V =   .0010000
DEFOCUS DISTANCE   DFD =   .000 mm

Figure 66 Spot diagram for lens MODEL1. Two spots are shown: without asymmetric fields (round spot); with asymmetric fields at the same focal plane (elliptical spot), showing the astigmatism introduced by the two slots and ports.

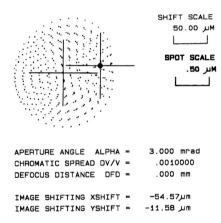

APERTURE ANGLE   ALPHA =   3.000 mrad
CHROMATIC SPREAD DV/V =   .0010000
DEFOCUS DISTANCE   DFD =   .000 mm

IMAGE SHIFTING XSHIFT =   −54.57μm
IMAGE SHIFTING YSHIFT =   −11.58 μm

Figure 67 Spot diagram for lens MODEL2. Two grids are shown: without asymmetric fields (thick grid, small spot); with asymmetric fields (thin grid, large spot), showing the spot deflection and beam blur due to off-axis slot and port.

heavy line is the grid and spot predicted if no asymmetric fields are present; the thin line is the grid and spot predicted when the asymmetric fields are taken into account.

In Fig. 66, there is no image shifting so the two grids overlap. The heavy dots at the intersections of the grid lines are the spot size and shape for a rotationally symmetric lens. As can be seen, they are round and about 0.06 μm in diameter. The spots in the asymmetric lens (with slots and ports) are also centered at the grid line intersects, but are about six times larger

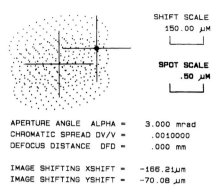

SHIFT SCALE
150.00 μM

SPOT SCALE
.50 μM

APERTURE ANGLE ALPHA =	3.000 mrad
CHROMATIC SPREAD DV/V =	.0010000
DEFOCUS DISTANCE DFD =	.000 mm
IMAGE SHIFTING XSHIFT =	-166.21μm
IMAGE SHIFTING YSHIFT =	-70.08 μm

Figure 68 Spot diagram for lens MODEL3. Two grids are shown: without asymmetric fields (thick grid, small spot); with asymmetric fields (thin grid, extended spot), showing a large image shift and beam blur.

and are not round. The effect of the wafer slots and evacuation ports (which are rotated by 90° with respect to one another) in MODEL1 is evident in the plot. The predicted spots have a distinctive fourfold symmetry due to the flux leakage in the two orthogonal directions.

The presence of the long, asymmetrically centered rectangular box in MODEL2 is reflected in the spot shapes and positions of Fig. 67. The whole image is shifted and the individual spots are flattened at the right and smeared out from left to right. The symmetry displayed by the spots of Fig. 66 has disappeared. As mentioned before, this image shift can be corrected by application of a D.C. offset to the deflection waveform.

Lens MODEL3 is, as one might expect, the worst optically. Its image is shifted, elongated, and seems to have a much higher chromatic aberration due to the stray fields than the other two lenses. The actual beam blur of lens MODEL3 for a monochromatic source is, however, about the same as that of lens MODEL2 (approximately 1.5 μm).

7.4 Summary

In conclusion, if the existing deflectors were adjusted to compensate for the image shift and a conventional stigmator were used to correct the astigmatism, the optical defects of lens MODEL3 could be sufficiently corrected to enable it to be a reasonable final lens for an inspection system (provided the energy spread in the source is small). In addition, the extra volume it allows for the positioning of stages and collectors, and its flat exterior surfaces means it is a better specimen chamber than the other models considered.

With other advantages, such as shielding and lower primary aberrations, this study has shown that the wafer-in-lens concept is viable for low energy inspection systems.

8. SYSTEMS WITH COMBINED 3D ELECTRIC AND MAGNETIC FIELDS

This section is devoted to two examples where both a 3D magnetic and a 3D electrostatic analysis are required, and the behavior of the electrons is affected by the combined presence of the fields. The two cases considered are only intended for the purpose of illustration and are not meant to be taken as design studies on systems which are in any way optimized or even practical in their present state. They do, however, serve to show where the capability of the new software might be invaluable from a designer's point of view.

The first example is of a Wien filter, which can be used to select electrons of a specified energy. It uses the fact that the transverse force on an electron due to crossed electric and magnetic fields can be canceled out if the fields are in the correct ratio relative to the electron's speed.

The second example is of a combined magnetic focussing and deflection system and an off-axis electrostatic secondary electron collector. Such a system is typical of those used in the final stages of an SEM. Three arrangements are presented, which highlight the need to use both a combined 3D electric and magnetic ray-trace and a 3D optical properties calculation program to be used during the design of such systems.

8.1 Analysis of a Wien Filter

Two defects which are prevalent in most electron optical systems are spherical and chromatic aberration. Spherical aberration cannot be completely eliminated by the use of round lenses, but in some cases, it can be reduced significantly by lowering the numerical aperture of the lens or increasing the lens strength so as to reduce its focal length. If this is done, the size of the image spot may then be limited by chromatic effects. This is especially true in low voltage applications where the electron interactions are more significant and the ripple in the power supplies is proportionally worse. It is also of particular importance in ion beam systems, where there is a large energy spread in the source. By limiting the energy spread in an electron or ion beam, it is possible to significantly reduce the chromatic spot size.

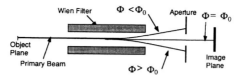

Figure 69 Illustration of the dispersion of charged particles by Wien filter according to their energy.

8.1.1 Principle of the Wien Filter

One particular monochromating filter is that introduced by Wien in 1898 (Wien, 1898) in which particles traveling along a straight axis pass through a region of crossed electric and magnetic fields. The fields are arranged so that they are mutually orthogonal and also orthogonal to the direction of motion of the charged particles. Under these conditions, the transverse force, F, on a particle of charge q, traveling at velocity v, will be:

$$F = q(E + vB)$$

where E is the electric field and B is the magnetic flux density, and this force acts normally to the direction of motion. The particles with a nominal speed v_0 will pass through the filter undeflected if $F = 0$, i.e., if $v_0 = -E/B$. The transverse force can, therefore, be written $F = qE(1 - v/v_0)$ or:

$$F = qE\left(1 - \sqrt{\frac{\Phi}{\Phi_0}}\right)$$

where Φ is the energy of the particle and Φ_0 is the nominal energy.

Particles with energies greater than Φ_0 will be deflected to one side of the axis and particles with energies less than Φ_0 will be deflected to the other side of the axis (i.e., the beam is dispersed according to the energies of the particles) – see Fig. 69. Particles with a limited range of energies close to the nominal energy, Φ_0, can be selected by passing the beam through an aperture after the filter.

8.1.2 Geometry of the Filter and Axial Fields

A Wien filter based on a design described by Tsuno (1990) was analyzed.

The structure of the magnetic part of the filter is shown in Fig. 70 in $(x$–$y)$ cross-section with magnetic scalar equipotential lines corresponding to 1 Ampere turn. The coils drive the flux around the outer magnetic

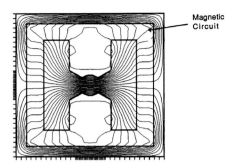

Figure 70 *x–y* cross-section of magnetic circuit of Wien filter with magnetic scalar equipotentials superposed.

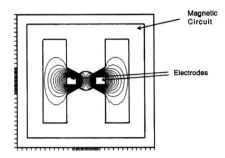

Figure 71 *x–y* cross-section of electrostatic circuit of Wien filter with electrostatic equipotentials superposed.

circuit and across the gap in the center. The length of the filter in the *z*-direction is 100 mm.

The electric part of the filter is shown in Fig. 71 in (*x–y*) cross-section with electrostatic equipotentials superposed. A voltage of ± 1 V was applied to the electrodes and the polepieces were grounded. Fig. 72 is a cut-away 3D view of the central part of the filter.

Fig. 73 shows the magnetic and electrostatic axial deflection fields. As can be seen they are almost (but not exactly) identical in shape. The central part of each curve is approximately flat. This corresponds to the part of the axis inside the filter. The deflection fringing fields near and beyond the ends of the filter do not exactly match, the electric field falling off more sharply than the magnetic. The electric field inside the filter is, $E_0 = 138.1$ V/m, and the magnetic flux density inside the filter is, $B_0 = 1.04 \times 10^{-3}$ T.

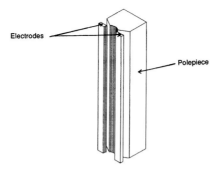

Figure 72 3D cut-away view of the central region of the Wien filter, showing the electrodes and the polepiece.

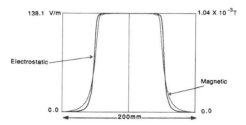

Figure 73 Computed electrostatic and magnetic axial deflection fields in the Wien filter, showing that the two are almost always matched, except in the fringing region.

8.1.3 Ray-Tracing of Electrons Through the Filter

The nominal energy (i.e., the kinetic energy of the undeflected electrons) was chosen to be $\Phi_0 = 1000$ eV. The nominal velocity v_0 is given by:

$$\frac{1}{2}mv_0^2 = e\Phi_0$$

where m is the mass of the electron and e its charge; i.e.,

$$v_0 = \sqrt{2\Phi_0\frac{e}{m}} = 1.876 \times 10^7 \text{ m s}^{-1}$$

Therefore, for 1 keV electrons to pass undeflected through the filter, the electric and magnetic fields must be in the ratio $E/B = v_0 = 1.876 \times 10^7$ m s^{-1}. Now, $E_0/B_0 = 1.33 \times 10^5$ m s^{-1} and therefore the electric field should be scaled up by a factor of $187.6/1.33 = 141.25$, relative to the magnetic field for the combined field ray-trace to allow 1 keV electrons to

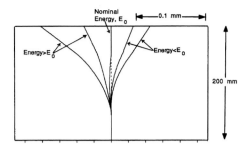

Figure 74 Trajectories of electrons with energies above, below, and at the nominal energy passing through the Wien filter at low excitation, showing that the nominal electrons are undeflected.

pass through undeflected. Fig. 74 shows electrons at 1200, 1100, 1000, 900, and 800 eV passing from bottom to top through the filter. The electric field used was $1.4125 \times E_0$ and the magnetic flux density used was $0.01 \times B_0$.

As can be seen, the slower electrons are deviated to the left and the faster ones to the right. Those at the nominal energy pass through the filter with no deviation at the exit plane.

The electrons with nominal energy are, however, slightly deflected as they pass through the filter. They move to the left on entering the filter and back onto the axis on leaving the filter. This is presumably due to the non-cancellation of the electric and magnetic deflection fields at certain places on the path of the beam (as shown on the axis in Fig. 73). The non-matching of the deflection fields will ultimately limit the sensitivity of the filter. This is because as the coil currents and electrode voltages are increased (and, therefore, the sensitivity), the excursions off the axis of the nominal electrons will become progressively greater where the fields do not cancel out. Fig. 75 shows the trajectories of electrons with the same initial energies as those in Fig. 74 but passing through stronger electric and magnetic fields in the filter: the electric field used was $14.125 \times E_0$ and the magnetic flux density used was $0.1 \times B_0$.

It can be seen that, in this case, the excursion of electrons at the nominal energy is significant. Matching the axial deflection fields at all points on the electron path is, therefore, crucial in the design of such filters, and the 3D effects near each end of the filter are an important consideration in the design.

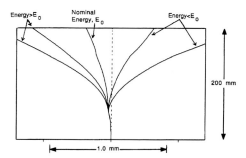

Figure 75 Trajectories of electrons with energies above, below, and at the nominal energy passing through the Wien filter in the high excitation case showing that the mismatch of the fields in the fringing region causes significant deflection even at the nominal energy when a high sensitivity is required.

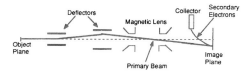

Figure 76 Schematic diagram of focussing and deflection system with off-axis electrostatic secondary electron collector.

8.2 3D Focussing, Deflection, and Collection System for an SEM

This section illustrates the interfacing between the 3D programs, the STRAYFIELD programs, and the LITHO (Chu & Munro, 1981) software package, which enables combinations of magnetic and electrostatic lenses and deflectors in the presence of 3D stray fields to be analyzed. As an example, consider the following setup (Fig. 76). A magnetic lens and deflection system is used to form an image which is scanned over a 5 mm field. There are two saddle type deflection coils placed before the lens gap. Between the lens and the image plane there is an off-axis, electrostatic secondary electron collector at a positive potential.

The magnetic lens is round but has a hole drilled through its sides normal to the optic axis. Fig. 76 shows a schematic of the system and Fig. 77 shows a cut-away view of the lens and detector.

The stray fields in this example come from two sources. Firstly there is an asymmetrical magnetic field arising from the hole drilled through the round magnetic lens. Secondly there is an electrostatic asymmetry due to the presence of the off-axis secondary electron collector.

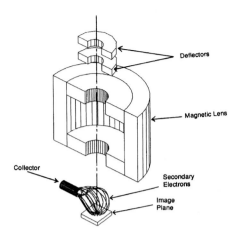

Figure 77 3D cutaway diagram of focussing and deflection system with off-axis electrostatic SE collector.

For the purposes of this example, the beam voltage is denoted by V_0, the collector potential V_c, the image position Z_i, and the excitation of the lens NI. The object position was fixed at $Z_o = 0$ mm, the deflectors at $z = 86$ mm and $z = 131$ mm, the lens at $z = 140$ mm and the collector at $Z_i - 30$ mm.

8.2.1 Arrangement 1: Good Collection

The image position was set at $Z_i = 300$ mm, the beam voltage was $V_0 = 1000$ V, the collector potential was $V_c = 200$ V, and the lens excitation required to focus the beam was $NI = 180$ AT. Fig. 78 is a 2D projection of a 3D ray-trace through the combined fields and, as can be seen, most of the emitted secondaries are collected. On the basis of this ray-trace, it appears that the configuration is suitable.

The 3D axial fields were then extracted using programs EAXIAL and MAXIAL (see Section 3.7) and used by the STRAYFIELD (Zhu, 1989) programs to compute the aberrations caused by the detector and the hole in the magnetic lens. Fig. 79 is a diagram of the spot, with and without the electric and magnetic stray fields. (It was found that the effect of the non-rotationally symmetric magnetic field perturbations was negligible compared with the effect of the electrostatic collector.) As this diagram shows, the optics of the configuration are horrendous. The spot size is of the order of 300 μm and the whole image is shifted diagonally by about

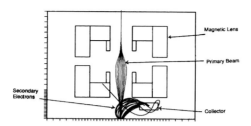

Figure 78 2D projection of a 3D ray-trace of secondary electrons leaving the sample for arrangement 1, showing that most of the emitted secondaries reach the detector.

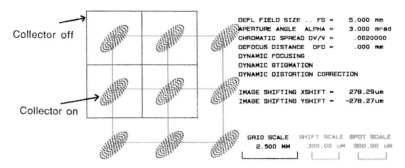

Figure 79 Spot diagram at the image plane for arrangement 1, showing that the collector field significantly impairs the optical performance of the system.

400 μm (the detector was rotated by 45° relative to the lens and deflectors). In addition, the spot is very elongated. On this basis, the system is clearly not suitable.

8.2.2 Arrangement 2: Good Optics

In this case, $Z_i = 300$ mm, $V_0 = 25$ kV, $V_c = 10$ V, and $NI = 1006$ AT. Fig. 80 shows the spot size to be about 0.5 μm and the image shift to be about 0.7 μm. Clearly this is an acceptable system from an optics point of view.

Fig. 81 is a 2D projection of a 3D ray-trace of secondaries leaving the specimen with initial energies of a few eVs. It shows that, unfortunately, none of the electrons reach the detector, since the specimen is now immersed in the magnetic field of the lens. From the point of view of secondary electron collection, the system is useless.

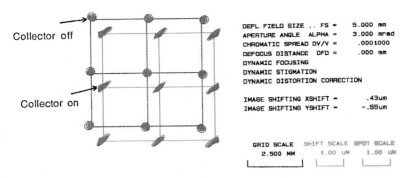

Figure 80 Spot diagram at the image plane for arrangement 2, showing that the spot degradation is minimal when the collector is switched on in this system.

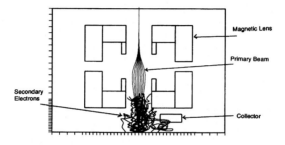

Figure 81 2D projection of a 3D ray-trace of secondary electrons leaving the sample for arrangement 2, showing that almost no signal is obtained in this case.

8.2.3 Arrangement 3: Good Collection, Good Optics

In this case, $Z_i = 400$ mm, $V_0 = 25$ kV, $V_c = 100$ V, and $NI = 900$ AT. Fig. 82 shows the spot size to be about a micrometer and the image shift to be about 0.1 μm and Fig. 83 shows that most of the electrons are collected. This configuration is therefore good from both a collection and an optics point of view.

8.2.4 Summary

The use of either the 3D combined field direct ray-trace or an optical properties program individually does not give a complete evaluation of the focussing and collection system. Using the two programs together enables a system to be designed which has good optical properties and good secondary electron collection.

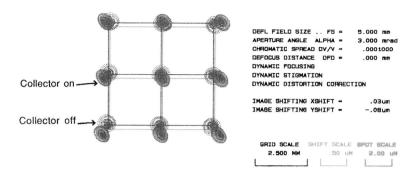

```
DEFL FIELD SIZE .. FS =      5.000 mm
APERTURE ANGLE   ALPHA =     3.000 mrad
CHROMATIC SPREAD DV/V =       .0001000
DEFOCUS DISTANCE  DFD =       .000 mm
DYNAMIC FOCUSING
DYNAMIC STIGMATION
DYNAMIC DISTORTION CORRECTION

IMAGE SHIFTING XSHIFT =        .03um
IMAGE SHIFTING YSHIFT =       -.08um


GRID SCALE   SHIFT SCALE   SPOT SCALE
   2.500 MM       .50 uM     2.00 uM
```

Figure 82 Spot diagram at the image plane for arrangement 3, showing that the collector field does not affect the primary beam optics significantly in this case.

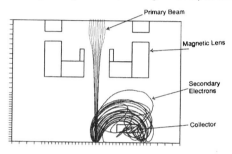

Figure 83 2D projection of a 3D ray-trace of secondary electrons leaving the sample for arrangement 3, showing a good collection efficiency for this system to match the good optical performance.

9. CONCLUSIONS

The new extensions to the finite difference method described enable generation of compact and unified computer code for the computation of fully three-dimensional electrostatic and magnetic field distributions in systems that include: electrodes of fixed potential; charged or uncharged dielectric materials; permeable ferromagnetic materials; and current carrying coils.

The new computer programs constitute a versatile tool for the design of electron and other charged particle beam devices. The software can accurately analyze complex 3D geometries on the latest generation of personal microcomputers, as the illustrative examples have shown. The unified format of the data and the integrated nature of the program elements for electrostatic and magnetic analyses enables convenient and self-contained modeling of most parts of the column.

The work described opens up the possibility of further research in two main areas: firstly, improvements and enhancements to the code to include, for example, the treatment of magnetic saturation and space charge; secondly, the analysis of novel 3D electron optical components, for example, multiple beam systems with cathode and lens arrays, multipole systems and their tolerancing, and advanced electron detectors for electron microscopes and inspection systems.

ACKNOWLEDGMENTS

I would like to thank my PhD supervisor, Dr. Eric Munro, and other colleagues and friends, especially Xieqing Zhu, Dr. Pat Thornton, Prof. Tom Mulvey, and Dr. Dan Meisburger for many helpful discussions. I am also indebted to my wife Sandra for her assistance with the artwork and all her support.

I gratefully acknowledge the financial support provided for this work by the Science and Engineering Research Council, KLA Instruments Inc., Hitachi Instruments Inc., and Siemens A.G.

REFERENCES

Allen, D. N. G. (1954). *Relaxation methods*. McGraw-Hill.

Chu, H. C., & Munro, E. (1981). *A set of computer programs for the design and optimisation of electron beam lithography systems*. Imperial College internal report.

Costabel, M. (1987). Principles of boundary element methods. In *Computer physics reports: Vol. 6* (pp. 243–274). Amsterdam: North-Holland Physics Publishing.

Desbruslais, S. R., & Munro, E. (1987). In J. Ximen (Ed.), *Proceedings of the international symposium on electron optics* (pp. 45–48). Beijing: Institute of Electronics, Academia Sinica.

Frankel, S. P. (1950). *Mathematical Tables and Other Aids to Computation, 4*, 65.

Franzen, N. (1984). In J. J. Hren, et al. (Eds.), *Electron optical systems for microscopy, microanalysis and microlithography* (pp. 115–126). Chicago: SEM Inc.

Imperial College electron optics software catalogue (1990). Advertising literature.

Jacobs, D. A. H. (1983). Preconditioned conjugate gradient algorithms for solving finite difference equations. In D. J. Evans (Ed.), *Topics in computational mathematics: Vol. 1*. New York: Gordon and Breach.

Joy, D. C. (1988). *Scanning Microscopy, 2*, 1901–1915.

Joy, D. C. (1989). *SEGEN*. Original PASCAL source code obtained from David Joy.

Kamminga, W., Verster, J. L., & Franken, J. C. (1968). *Optik, 28*, 442.

Kasper, E. (1982). Magnetic field calculations and determination of trajectories. In P. W. Hawkes (Ed.), *Magnetic electron lenses* (pp. 57–118). Berlin: Springer-Verlag.

Khursheed, A., & Dinnis, A. R. (1989). *Journal of Vacuum Science & Technology B, 7*, 1882–1885.

Lencova, B., & Lenc, M. (1984). *Optik, 68*, 34.

Lencova, B., & Wisselink, G. (1990). *Nuclear Instruments & Methods in Physics Research, Section A, 298*, 56–66.

MacGregor, D. M. (1983). In *IEEE conference proceedings of the international conference on consumer electronics* (pp. 130–137).

Meijerink, J. A., & Van der Vorst, H. A. (1977). *Mathematics of Computation, 31*, 148–167.

Morton, K. W. (1987). Basic course in finite element methods. In *Computer physics reports: Vol. 6* (pp. 1–72). Amsterdam: North-Holland Physics Publishing.

Munro, E. (1971). *Computer-aided-design methods in electron optics* (PhD thesis). University of Cambridge.

Munro, E. (1973). In P. W. Hawkes (Ed.), *Image processing and computer aided design in electron optics* (pp. 284–323). London: Academic Press.

Munro, E. (1975). *A set of programs for calculating the properties of electron lenses.* Cambridge University Engineering Department Report CUED/B-ELECT/TR 45.

Munro, E. (1988). *Journal of Vacuum Science & Technology A, 6,* 941–948.

Munro, E. (1990). *Journal of Vacuum Science & Technology A, 8,* 1657–1665.

Munro, E., & Chu, H. C. (1982). *Optik, 61,* 1.

Nunn, J. (1991). PhD thesis. Imperial College, University of London.

Plies, E. (1990). *Nuclear Instruments & Methods in Physics Research, Section A, 298,* 142–155.

Shao, Z., & Lin, P. S. D. (1989). *Review of Scientific Instruments, 11,* 3434.

Sheppard, W. F. (1899). *Proceedings of the London Mathematical Society, 31,* 441–488.

Silvester, P. P., & Ferrari, R. L. (1990). *Finite elements for electrical engineers* (2nd edition). Cambridge.

Simpkin, J., & Trowbridge, C. W. (1979). *International Journal for Numerical Methods in Engineering, 14,* 423–440.

Smith, M. R., & Munro, E. (1987). *Journal of Vacuum Science & Technology B, 5,* 161.

Southwell, R. V. (1940). *Relaxation methods in engineering science.* Oxford University Press.

Southwell, R. V. (1946). *Relaxation methods in theoretical physics.* Oxford University Press.

Sturrock, P. A. (1951). *Philosophical Transactions of the Royal Society of London, 243,* 387.

Tsuno, K. (1990). *Nuclear Instruments & Methods in Physics Research, Section A, 298,* 296–320.

Weber, C. (1967). Numerical solution of Laplace's and Poisson's equations and the calculation of electron trajectories and electron beams. In A. Septier (Ed.), *Focusing of charged particles: Vol. 1* (pp. 45–99). New York: Academic Press (Chapter 1.2).

Wien, W. (1898). *Verhandlungen der Deutschen Physikalischen Gesellschaft, 16,* 165–172.

Young, D. M. (1954). *Transactions of the American Mathematical Society, 76,* 92–111.

Zhu, X. (1989). *A computer program for computing asymmetry aberrations for systems in the presence of 3D stray fields.* Private communication.

Zhu, X., & Munro, E. (1989). *Journal of Vacuum Science & Technology B, 7,* 1862–1869.

FURTHER READING

In addition to references listed above, which were made at specific points in the section, the following books and articles were a constant source of reference during the work:

Boas, M. L. (1983). *Mathematical methods in the physical sciences* (2nd edition). Wiley.

Hawkes, P. W., & Kasper, E. (1989). *Principles of electron optics: Vols. 1 and 2.* London: Academic Press.

Klemperer, O., & Barnett, M. E. (1971). *Electron optics.* Cambridge University Press.

Munro, E. (1980). *Electron optics MSc course notes.* London: Applied Optics Section, Imperial College.

Paddon, D. J., & Holstein, H. (Eds.). (1985). *Multigrid methods for integral and differential equations.* Oxford: Clarendon Press.

Press, W. H., Flannery, B. P., Tuekolsky, S. A., & Vetterling, W. T. (1986). *Numerical recipes.* Cambridge.

Smythe, W. R. (1968). *Static and dynamic electricity* (3rd edition). McGraw-Hill.

INDEX

Printed in the United States
By Bookmasters